Electrical and Instrumentation Engineering

Electrical and Instrumentation Engineering

Zain Jackson

New York

Published by NY Research Press
118-35 Queens Blvd., Suite 400,
Forest Hills, NY 11375, USA
www.nyresearchpress.com

Electrical and Instrumentation Engineering
Zain Jackson

International Standard Book Number: 978-1-64725-423-0 (Hardback)

Cataloging-in-Publication Data

Electrical and instrumentation engineering / Zain Jackson.
   p. cm.
Includes bibliographical references and index.
ISBN 978-1-64725-423-0
1. Electrical engineering. 2. Engineering instruments. 3. Electric apparatus and appliances.
4. Engineering. I. Jackson, Zain.
TK145 .E44 2023
621.3--dc23

Contents

Preface

Electrical engineering refers to a branch of engineering that focuses on the design, research, and use of systems, machinery, and gadgets that rely on electromagnetism, electricity, and electronics. Instrumentation engineering is the science of measuring and controlling process variables in an industrial or production area. Electrical and instrumentation engineering (EIE) is a subfield of electrical engineering that deals with managing equipment for automated control and measuring process variables inside industrial facilities. It is concerned with the design of instruments that measure physical quantities like flow, temperature, and pressure. This book contains a detailed explanation of the various concepts and applications of electrical and instrumentation engineering. While understanding the long-term perspectives of the topics, it makes an effort in highlighting their impact as a modern tool for the growth of the discipline. This book is an essential guide for both academicians and those who wish to pursue this discipline further.

This book has been the outcome of endless efforts put in by authors and researchers on various issues and topics within the field. The book is a comprehensive collection of significant researches that are addressed in a variety of chapters. It will surely enhance the knowledge of the field among readers across the globe.

It gives us an immense pleasure to thank our researchers and authors for their efforts to submit their piece of writing before the deadlines. Finally in the end, I would like to thank my family and colleagues who have been a great source of inspiration and support.

Zain Jackson

Electrical Circuits

1.1   Basic Circuit Components

The three basic circuit components are as follows:

- Resistor
- Capacitor
- Inductor

Resistor

It is defined as the property of the material by which it opposes the flow of current through it:

- It is denoted by R whose unit is Ω.

Symbol

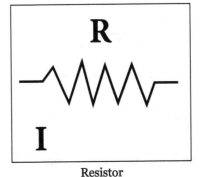

Resistor

The relation between V and I is given by Ohm's law as follows,

$$V = IR$$

Capacitor

In both the digital and analog electronic circuits, a capacitor is a fundamental element. It filters the signals and provides a fundamental memory element.

The capacitor is an element which stores energy in the electric field. The circuit symbol and associated electrical variables for the capacitor is shown in the below figure:

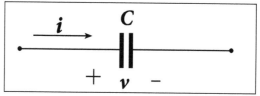

Circuit symbol for capacitor.

The capacitor is modeled as two conducting plates which are separated by a dielectric as shown in the above figure. When a voltage v is applied across the plates, a charge +q accumulates on one plate and a charge - q accumulates on the other plate.

If the plates have an area A which are separated by a distance d, the electric field generated across the plates is given as,

$$E = \frac{q}{\varepsilon A} \qquad ...(1)$$

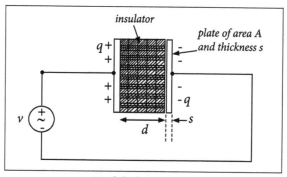

Model of capacitor.

The voltage across the capacitor plates is given as,

$$v = Ed = \frac{qd}{\varepsilon A} \qquad ...(2)$$

Current flowing into the capacitor is the rate of change of charge across the capacitor plates which is given as,

$$i = \frac{dq}{dt}$$

$$i = \frac{dq}{dt} = \frac{d}{dt}\left(\frac{\varepsilon A}{d}v\right) = \frac{\varepsilon A}{d}\frac{dv}{dt} = C\frac{dv}{dt} \qquad ...(3)$$

The constant of proportionality(C) is known as the capacitance of the capacitor.

The capacitor has a plate separation (d) and plate area (A) where ε is the permittivity of the dielectric material between the plates,

$$C = \frac{\varepsilon A}{d} \qquad ...(4)$$

Capacitance represents the efficiency of the charge stored and it is measured in units of Farads (F).

Current-voltage relationship of the capacitor is given as,

$$i = C\frac{dv}{dt} \qquad ...(5)$$

For DC signals (dv/dt = 0), the capacitor acts as an open circuit (i = 0). The capacitor does not like voltage discontinuities since the current at infinity is not possible.

If we integrate equation (5) over time, we have,

$$\int_{-\infty}^{t} i\,dt = \int_{-\infty}^{t} C\frac{dv}{dt}\,dt \qquad ....(6)$$

$$v = \frac{1}{C}\int_{-\infty}^{t} i\,dt$$

$$= \frac{1}{C}\int_{0}^{t} i\,dt + v(0) \qquad ...(7)$$

The constant of integration v(0) represents the voltage of the capacitor at time t = 0. The presence of the constant of integration v(0) is the reason for the memory properties of the capacitor.

Let us consider the circuit shown in the below figure where the capacitor of capacitance C is connected with a time varying voltage source v(t).

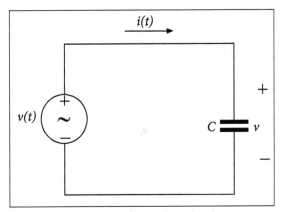

Fundamental capacitor circuit.

The voltage v(t) is given as,

$$v(t) = A \cos(\omega t) \qquad ...(8)$$

Then the current i(t) is given as,

$$i(t) = C \frac{dv}{dt}$$

$$= -CA \, \omega \sin(\omega t)$$

$$= C \omega A \cos\left(\omega t + \frac{\pi}{2}\right) \qquad ...(9)$$

Hence, the current going through the capacitor and the voltage across the capacitor are 90° out of phase. It is said that the current leads the voltage by 90°. The general plot of the voltage and the current of the capacitor is shown in the below figure. The current leads the voltage by 90°.

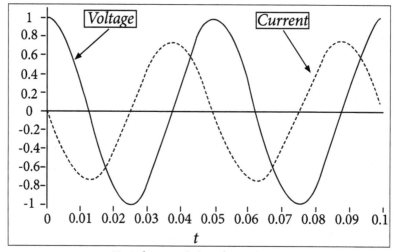

V - I characteristics of capacitors.

If we take the ratio of the peak voltage to the peak current, we have,

$$X_C = \frac{1}{C\omega} \qquad ...(10)$$

The unit of X_c is Volts/Amperes or Ohms which represents resistance. As the frequency $\omega \to 0$, the quantity X_c goes to infinity which implies that the capacitor resembles an open circuit.

As the frequency becomes very large $\omega \to \infty$, the quantity X_c goes to zero which implies that the capacitor resembles a short circuit.

Inductor

The inductor is a coil which stores the energy in the magnetic field. Let us consider a wire of length l forming a loop of area A as shown in the below figure. A current i(t) is flowing through the wire as indicated in the below figure. This current generates the magnetic field B which is given as,

$$B(t) = \mu\, i(t)/l \qquad ...(1)$$

Where,

μ = Magnetic permeability of the material enclosed by the wire

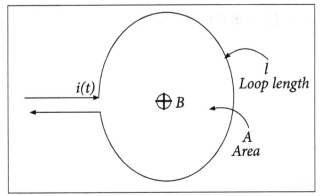

Current loop for the calculation of inductance.

The magnetic flux ϕ through the loop of the area A is denoted as,

$$\phi = AB(t)$$

$$= A\,\mu\, i(t)/l$$

$$L = A\mu/l \qquad ...(2)$$

From Maxwell's equations, we have,

$$\frac{d\Phi}{dt} = v(t) \qquad ...(3)$$

$$\frac{d\,Li(t)}{dt} = v(t) \qquad ...(4)$$

$$v = L\frac{di}{dt} \qquad ...(5)$$

The parameter L is known as the inductance of the inductor. Its unit is Henry (H).

The circuit symbol and associated electrical variables for the inductor is shown in the below figure:

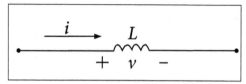

Circuit symbol of inductor.

For DC signals ($di/dt = 0$), the inductor acts as a short circuit ($v = 0$). The inductor does not like current discontinuities since the voltage across it becomes infinity which is not physically possible.

If we integrate equation (5) over time, we have,

$$\int_{-\infty}^{t} v \, dt = \int_{-\infty}^{t} L \frac{di}{dt} dt \qquad \ldots(6)$$

$$i = \frac{1}{L} \int_{-\infty}^{t} v \, dt$$

$$= \frac{1}{L} \int_{0}^{t} v \, dt + i(0) \qquad \ldots(7)$$

The constant $i(0)$ represents the current through the inductor at time $t = 0$. The current at $t = -\infty$ is zero.

Let us now consider the circuit shown in the below figure where an inductor of inductance L is connected to a time varying current source $i(t)$.

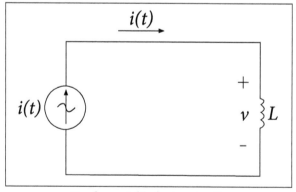

Fundamental inductor circuit.

The current $i(t)$ is given as,

$$i(t) = I_0 \cos(\omega t) \qquad \ldots(8)$$

The voltage v(t) becomes,

$$v(t) = L\frac{di}{dt}$$

$$= -LI_o \omega \sin(\omega t)$$

$$= L\omega I_o \cos\left(\omega t + \frac{\pi}{2}\right) \qquad \ldots(9)$$

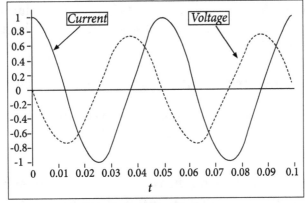

V-I relationship of inductor.

Thus, the current going through an inductor and the voltage across the inductor are 90° out of phase. Here the voltage leads the current by 90°. The general plot of the voltage and current of an inductor is shown in the above figure.

1.2   Ohm's Law and Kirchhoff's Law

Ohm's Law

It states that the current through a conductor between two points is directly proportional to the potential difference or voltage across the two points and inversely proportional to the resistance between them.

$$I = V/R$$

Where,

I - Current through the resistance in units of amperes.

V - Potential difference measured across the resistance in units of volts.

R - Resistance of the conductor in units of ohms.

Ohm's law states that R is constant and it is independent of the current.

Limitations of Ohm's law:

- It is not applicable to non-linear devices such as diodes, Zener diode, voltage regulators etc.

- Ohm's law is unsuitable for non-metallic conductors such as silicon carbide.

Kirchhoff's Law

Kirchhoff's laws are the basic analytical tools to obtain the solutions of voltages and currents for any electric circuit whether it is supplied from a direct current system or an alternating current system.

Kirchhoff's First Law or Kirchhoff's Current Law (KCL)

The total charge or current entering a junction or node is exactly equal to the charge leaving the node as it has no other place to go except to leave as no charge is lost within the node.

KCL states that, at any node (junction) in a circuit, the algebraic sum of currents entering and leaving a node at any instant of time must be equal to zero,

$$I_{(exiting)} + I_{(entering)} = 0$$

This idea by Kirchhoff is known as the conservation of charge.

At any instant of time, the algebraic sum of current at a junction or node is zero. Otherwise, the algebraic sum of current entering into a junction is equal to the current leaving the junction.

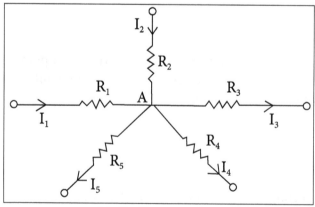

Kirchhoff's current law.

$$I_1 + I_2 = I_3 + I_4 + I_5$$

$$I_1 + I_2 - I_3 - I_4 - I_5 = 0$$

Kirchhoff's Second Law or Kirchhoff's Voltage Law (KVL)

In any closed loop network, the total voltage around the loop is equal to the sum of all the voltage drops within the same loop which is also equal to zero.

KVL states that in a closed circuit, the algebraic sum of all the source voltages must be equal to the algebric sum of all the voltage drops. This idea by Kirchhoff is known as the conservation of energy.

The algebric sum of potential around a closed path is equal to zero. In other words, the sum of potential rise is equal to the sum of potential drop in a closed circuit.

$$V = V_1 + V_2 + V_3$$

$$V - V_1 - V_2 - V_3 = 0$$

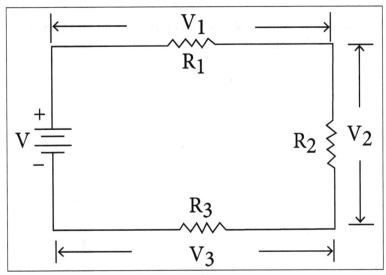

Kirchhoff's voltage law.

Steps to solve a problem using KCL/KVL method:

- Step 1: Find the branch currents of each branch using KCL.

- Step 2: Using KVL, get the equations for each loop in terms of unknown currents.

- Step 3: Solve the simultaneous equations obtained in step 2 using calculator to get the unknown currents.

- Step 4: Find the required branch current and power if required.

Problems

1. Let us consider a current of 0.55A is flowing through the resistance of 10Ω. Let us determine the potential difference between its ends.

Solution:

Given:

> Current, $I = 0.55A$
>
> Resistance, $R = 10\Omega$

Formula to be used:

> $V = IR$
>
> $= 0.55 \times 10$
>
> $= 5.5 \ V$

2. Let us consider a supply voltage of 200V is applied to a 100 Ω resistor. Let us determine the current flowing through it.

Solution:

Given:

> Voltage, V   200V
>
> Resistance, $R = 100\Omega$

Formula to be used:

> Current, $I = V / R$
>
> $= 200 / 100$
>
> $= 2 \ A$

3. Let us calculate the resistance of the conductor if a current of 2A flows through it when the potential difference across its ends is 12V.

Solution:

Given:

> Current, $I = 2A$
>
> Potential difference, $V = 12$

Formula to be used:

Resistance, $R = V / I$

$= 12 / 2$

$= 6$ ohm

4. Let us determine the current flowing in the 40Ω resistor R_3.

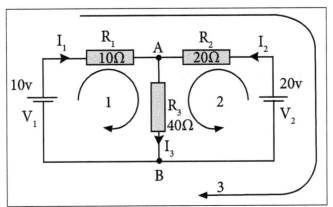

Solution:

The circuit has three branches with two nodes (A and B) and two independent loops.

Formula to be used:

Using Kirchhoff's current law (KCL), the equations are given as,

At node A,

$$I_1 + I_2 = I_3$$

At node B,

$$I_3 = I_1 + I_2$$

Using Kirchhoff's voltage law (KVL), the equations are given as,

Loop 1 is given as,

$$10 = R_1 \times I_1 + R_3 \times I_3 = 10I_1 + 40I_3 \qquad ...(1)$$

Loop 2 is given as,

$$20 = R_2 \times I_2 + R_3 \times I_3 = 20I_2 + 40I_3 \qquad ...(2)$$

Loop 3 is given as,

$$10-20 = 10I_1 - 20I_2 \quad ...(3)$$

I_3 is the sum of $I_1 + I_2$

We can rewrite the equations as,

Equation (1),

$$10 = 10I_1 + 40(I_1 + I_2) = 50I_1 + 40I_2$$

Equation (2),

$$20 = 20I_1 + 40(I_1 + I_2) = 40I_1 + 60I_2$$

We now have two simultaneous equations that can be reduced to give us the value of both I_1 and I_2.

Substitution of I_1 in terms of I_2 gives us the value of I_1 as -0.143 A.

Substitution of I_2 in terms of I_1 gives us the value of I_2 as +0.429 A.

$$I_3 = I_1 + I_2$$

Current flowing in resistor R_3 is given as,

$$-0.143 + 0.429 = 0.286A$$

Voltage across the resistor R_3 is given as,

$$0.286 \times 40 = 11.44 \text{ volts}$$

5. Let us determine the current through a 20Ω resistance and current through a 40Ω resistance.

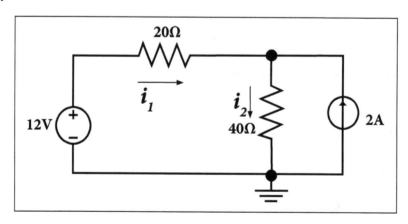

Solution:

By applying KCL at node x, we have,

$$-i_1 + i_2 - 2A = 0$$

v_x in the circuit using Ohm's law is given as,

$$i_1 = \frac{12V - v_x}{20\Omega}, i_2 = \frac{v_x}{40\Omega}$$

By applying last two equation into KCL at node x, we have,

$$-i_1 + i_2 - 2A = -\frac{12V - v_x}{20\Omega} + \frac{v_x}{40\Omega} - 2A = 0$$

$$-0.6 + 0.05\tilde{v}_x + 0.025\tilde{v}_x - 2 = 0$$

$$\tilde{v}_x = 34.67V$$

Current through a 20Ω resistance is given as,

$$i_1 = \frac{12V - 34.67}{20\Omega} = -1.133A$$

Current through a 40Ω resistance is given as,

$$i_2 = \frac{\tilde{v}_x}{40\Omega} = \frac{34.67}{40\Omega} = 0.866A$$

6. Let us find the current in a circuit using Kirchhoff's voltage law.

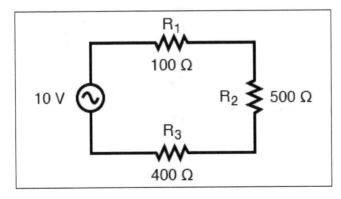

Solution:

$$10 = 100(I) + 500(I) + 400(I)$$

$$10 = 1000\left(I\right)$$

$$I = 10/1000$$

$$I = 0.01A$$

7. Let us determine the current i and voltage v over each resistor.

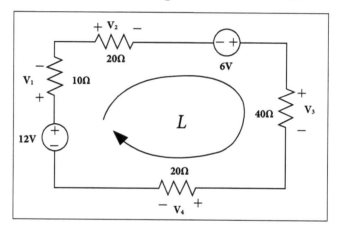

Solution:

KVL equations for voltages is given as,

$$V_1 + V_2 + V_3 + V_4 = 18$$

Using Ohm's law, we have,

$$V_1 = 10\Omega$$

$$V_2 = 20\Omega$$

$$V_3 = 40\Omega$$

$$V_4 = 20\Omega$$

$$10i + 20i + 40i + 20i = 18A$$

$$90i = 18A$$

$$i = \frac{18}{90} = 0.2\,A$$

$$V_1 = R_1 i = 10\left(0.2\right) = 2V$$

$$v_2 = R_2i = 20(0.2) = 4V$$

$$v_3 = R_3i = 40(0.2) = 8V$$

$$v_4 = R_4i = 20(0.2) = 4V$$

8. Let us determine V_1 and V_2 in the following circuit.

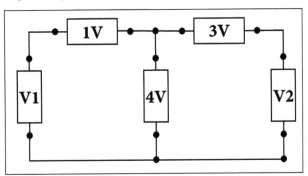

Solution:

Loop 1,

$$-V_1 + 1 + 4 = 0$$

$$V_1 = 5V$$

Loop 2,

$$-4 + 3 + V_2 = 0$$

$$V_2 = 1V$$

9. Let us determine V_1, V_2 and V_3 where the arrows are signifying the positive position of the box and the negative is at the end of the box.

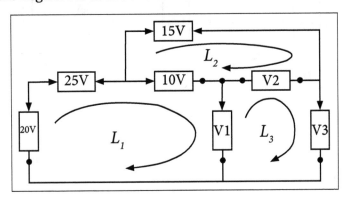

Solution:

Loop 1,

$$-20-25+10+V_1 = 0$$

$$V_1 = 35V$$

Loop 2,

$$-10+15-V_2 = 0$$

$$V_2 = 5V$$

Loop 3,

$$-V_1+V_2+V_3 = 0$$

$$-35+5+V_3 = 0$$

$$V_3 = 30V$$

10. Let us determine V_1, V_2, V_3 and V_4 from the below figure in which the arrows are signifying the positive position of the box and the negative is at the end of the box.

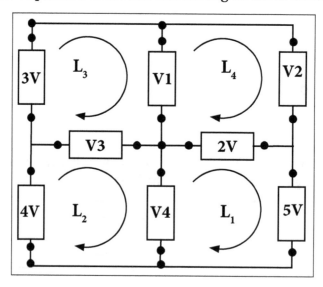

Solution:

Loop 1,

$$-V_4+2+5=0$$

$$V_4 = 7V$$

Loop 2,

$$4 + V_3 + V_4 = 0$$

$$V_3 = -4 - 7V$$

$$V_3 = -11V$$

Loop 3,

$$-3 + V_1 - V_3 = 0$$

$$V_1 = V_3 + 3$$

$$V_1 = -8V$$

Loop 4,

$$-V_1 - V_2 - 2 = 0$$

$$V_2 = -V_1 - 2$$

$$V_2 = 6V$$

11. Let us determine V_1, V_2 and V_3 in the following circuit.

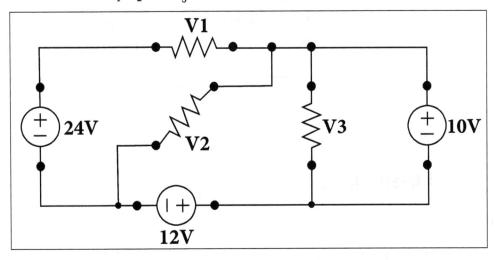

Solution:

Outer loop,

$$-24 + V_1 + 10 + 12 = 0$$

$$V_1 = 2V$$

V_2 and 10V source loop,

$$V_2 + 10 + 12 = 0$$

$$V_2 = -22V$$

V_3 and 10V source loop,

$$-V_3 + 10 = 0$$

$$V_3 = 10V$$

12. Let us determine I_1, I_2, I_3 in the following circuit.

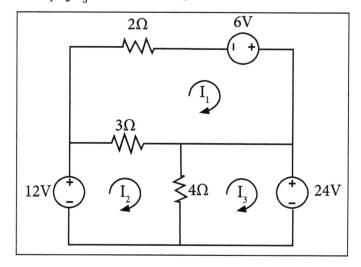

Solution:

Loop 1,

$$-6 = -2\,I_1 + 3\left(I_2 - I_1\right)$$

Loop 2,

$$12 = 3\left(I_2 - I_1\right) + 4\left(I_2 - I_3\right)$$

Loop 3,

$$24 = -4\left(I_3 - I_2\right)$$

From loop 3, we have,

$$\left(I_2 - I_3\right) = 6$$

From loop 2, we have,

$$12 = 3\left(I_2 - I_1\right) + 4(6)$$

$$I_2 - I_1 = -4$$

From loop 1, we have,

$$-6 = -2I_1 + 3(-4)$$

$$I_1 = -3A$$

From loop 2, we have,

$$I_2 - (-3A) = -4A$$

$$I_2 = -7A$$

From loop 3, we have,

$$24 = -4\ (I_3 + 7)$$

$$-6 = I_3 + 7$$

$$I_3 = -13A$$

1.3   Instantaneous Power

When a linear electric circuit is excited by a sinusoidal source, all currents and voltages in the circuit are also sinusoids of the same frequency as that of the excitation source. The below figure shows the general form of a linear AC circuit. The most general expressions for the voltage and current delivered to an arbitrary load are given as,

$$v(t) = V \cos(\omega t - \theta_V)$$

$$i(t) = I\cos(\omega t - \theta_I) \qquad \dots(1)$$

Where,

I and V = Peak amplitudes of the sinusoidal current and voltage.

θ_V and θ_I = Phase angles.

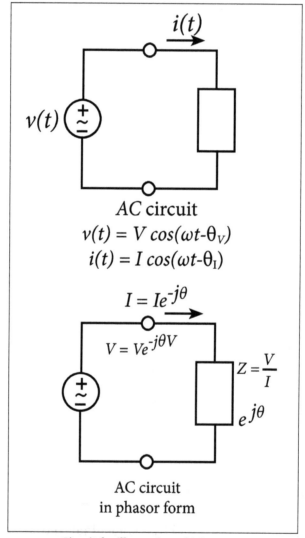

Circuit for illustration of AC power.

Two such waveforms are plotted in the below figure with unit amplitude and phase angles as $\theta_V = \pi/6$ and $\theta_I = \pi/3$.

The phase shift between the source and load is given as,

$$\theta = \theta_V - \theta_i$$

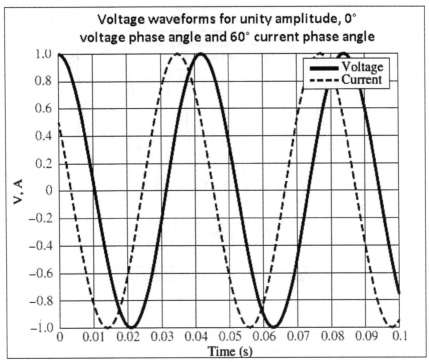

Current and voltage waveforms for illustration of AC power.

Since the instantaneous power dissipated by a circuit element is the product of the instantaneous voltage and current, then the power dissipated by an AC circuit element is given as,

$$p(t) = v(t)i(t) = VI \cos(\omega t) \cos(\omega t - \theta) \qquad ...(2)$$

$$p(t) = \frac{VI}{2} \cos(\theta) + \frac{VI}{2} \cos(2\omega t - \theta) \qquad ...(3)$$

Where,

θ = Difference in phase between voltage and current.

The above equation (3) illustrates the instantaneous power dissipated by an AC circuit element which is equal to the sum of an average component $1/2$ VI cos (θ) and a sinusoidal component $1/2$ VI cos($2\omega t$ - θ) oscillating at a frequency which is double that of the original source frequency. The instantaneous and average power for the signals of above figure are plotted in the below figure.

The average power corresponding to the voltage and current signals of equation (1) is obtained by integrating the instantaneous power over one cycle of the sinusoidal signal. Let T = $2\pi/\omega$ represent one cycle of the sinusoidal signals. Then the average power P_{av} is the integral of the instantaneous power p(t).

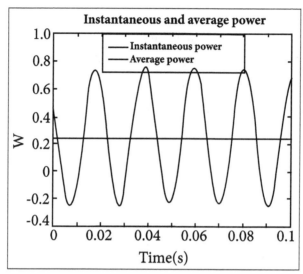

Instantaneous and average power dissipation.

Over one cycle, the average power given as,

$$P_{av} = \frac{1}{T}\int_0^T p(t)dt \qquad \dots(4)$$

$$= \frac{1}{T}\int_0^T \frac{VI}{0}\cos(\theta)dt + \frac{1}{T}\int_0^T \frac{VI}{2}\cos(2\omega t - \theta)dt$$

$$P_{av} = \frac{VI}{2}\cos(\theta) \qquad \dots(5)$$

The second integral is equal to zero and cos θ is a constant.

Same analysis carried out in equations (1) to (3) is repeated using phasor analysis. In phasor notation, the current and voltage of equation (1) is given as,

$$V(j\omega) = V\, e^{j0} \qquad \dots(6)$$

$$I(j\omega) = I\, e^{-j\theta} \qquad \dots(7)$$

$$Z = \frac{V}{I}e^{j(\theta)} = |Z|e^{j(\theta)} \qquad \dots(8)$$

The expression for the average power obtained from equation (4) represented using phasor notation is given as,

$$P_{av} = \frac{1}{2}\frac{V^2}{|Z|}\cos\theta = \frac{1}{2}I^2|Z|\cos\theta \qquad \dots(9)$$

1.4   Inductors and Capacitors

Inductive Networks

Inductance is the property of a coil by which it opposes the changes of current through it. It is referred as the electrical inertia possessed by the coil. The unit of inductance is Henry (L).

Inductance in denoted as L and it is given as,

$$L = \frac{\Psi}{I}$$

Where,

ψ - Total flux linkages produced by a current of I Amperes and $\psi = N\varphi$

$$L = \frac{N\phi}{I}$$

Where,

N - Number of turns of the coil.

ϕ - Flux linking with turns (Weber).

When a coil carries a changing current, an emf is induced in it due to the change of flux linkages. The induced emf is given as,

$$e = \frac{-L\, di}{dt}$$

Where,

L - Inductance,

di/dt - Rate of change of current with respect to time.

An inductor is a coil with number of turns and it posses inductance. An ideal inductor has zero ohmic resistance, whereas, a practical inductor has both inductance and resistance.

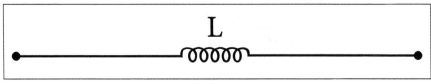

Inductor

Practical inductor.

When an inductor is used as a circuit element, a voltage develops across it when the current changes through it. When a current of constant magnitude flows, the inductor acts as short circuit and there is no voltage across it.

$$V = L\frac{di}{dt}$$

$$e = -L\frac{di}{dt}$$

$$I = \frac{1}{L}\int v\,dt$$

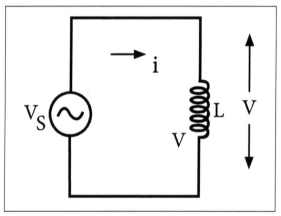

Induced emf.

Inductors in Series

Inductors in series are inductors that are placed back-to-back.

Below is a circuit where 3 inductors are placed in series:

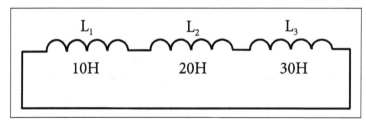

We can see the inductors are in series because they are back-to-back against each other. The best way to think of a series circuit is that if current flows through the circuit, the current can only take one path. We can see in the above circuit that if current flowed through it, it could only take one path.

Formula for Adding Inductors in Series

The formula to calculate the total series inductance of a circuit is,

$$L_T = L_1 + L_2 + L_3 + ...$$

So to calculate the total inductance of the circuit above, the total inductance, L_T would be,

$$L_T = 10H + 20H + 30H = 60H$$

So using the above formula, the total inductance is 60H.

Note: When inductors are in series, as the formula shows, they simply add together. Thus, the total inductance of a series circuit will always be greater than any of the individual inductor values.

Inductors in Parallel

Inductors in parallel are inductors that are connected side-by-side in different branches of a circuit.

Below is a circuit where 3 inductors are in parallel,

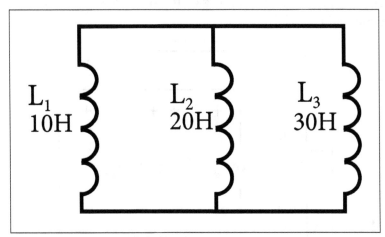

We can see that the inductors are in parallel because they are all on their own separate branches in the circuit. The best way to think about parallel circuits is by thinking of the path that current can take. When current is travelling through a parallel circuit, the current can take various paths through the circuits, such as to go through any of the branches containing the inductors. In series, this is not the case. Current can only take one path.

Formula for Adding Inductors in Parallel

The formula to calculate the total parallel inductance is,

$$L_T = \frac{1}{\frac{1}{L1} + \frac{1}{L2} + \frac{1}{L3} + ... + etc.}$$

So to calculate the total inductance of the circuit above, the total inductance, L_T would be,

$$L_T = \frac{1}{\frac{1}{10H} + \frac{1}{20H} + \frac{1}{30H}} = 5.45H$$

So using the above formula, the total inductance is 5.45Ω.

Note- When inductors are in parallel, the total inductance value is always less than the smallest inductor of the circuit. In other words, when inductors are in parallel, the total inductance shrinks. It's always less than any of the values of the inductors.

Capacitive Networks

Two conducting surfaces separated by a dielectric material constitute a capacitor. Dielectric material can be a solid, liquid or gas. It stores energy and acts like an insulator which cannot conduct electricity.

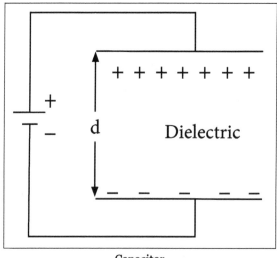

Capacitor

Dielectric strength is the maximum electric field intensity the material can bear without losing its insulating properties. When the stress exceeds its dielectric strength, it becomes a conductor. The electric stress is defined as force/unit charge or volts/m.

$$\text{Electric field intensity in the capacitor} = \frac{\text{voltage across capacitor}}{\text{separation between the plate}} = \frac{V}{d}.$$

Electric stress is expressed in kv/cm. An insulating material can withstand per cm length of the material without loosing their electrical properties. Air is an insulator and it can withstand 30 kv/cm at standard temperature and pressure (STP). Electric stress is used in the design of insulation for capacitor, cables, etc.:

- A capacitor possesses the inherent property of storing electric charges on its plate.

- When equal and opposite charges accumulate on the plates of a capacitor, a voltage develops across the capacitor.

- The charge required to raise the potential by one volt is termed as the capacitance of the capacitor.

Capacitance is given as,

$$C = \frac{Q}{V_C}$$

Where,

Q = Charge on the plates of the capacitor

The unit of capacitance is Farad.

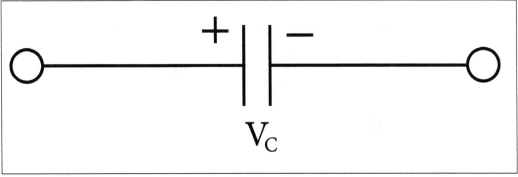

Symbol of capacitor.

If the capacitor current is time varying, then,

$$C = \frac{q}{V}$$

Where,

q = Instantaneous value of the charge

V = Instantaneous value of the voltage

Let i be the current through the capacitor

$$i = \frac{dq}{dt} = \frac{d}{dt}(CV) = C\frac{dV}{dt}$$

Capacitor current is given as,

$$i = C\frac{dV}{dt}$$

Voltage across the capacitor in terms of capacitor current is given as,

$$v(t) = \int \frac{idt}{C}$$

$$V = \frac{1}{C}\int i\,dt$$

Capacitors in Series

Capacitors in series means two or more capacitors connected in a single line. Positive plate of the one capacitor is connected to the negative plate of the next capacitor.

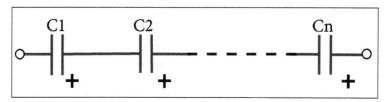

Here,

$$QT = Q_1 = Q_2 = Q_3 = ----- = Q$$

$$IC = I_1 = I_2 = I_3 = ---- = IN$$

When the capacitors are connected in series Charge and current is same on all the capacitors.

For series capacitors same quantity of electrons will flow through each capacitor because the charge on each plate is coming from the adjacent plate. So, coulomb charge is same. As current is nothing but flow of electrons, current is also same.

Equivalent Capacitance

Equivalent capacitance is the overall capacitance of the capacitors. Let us see how to calculate the capacitance when they are in series.

Below is the figure showing three capacitors connected in series to the battery. When the capacitors are connected in series the adjacent plates get charged due to electrostatic induction.

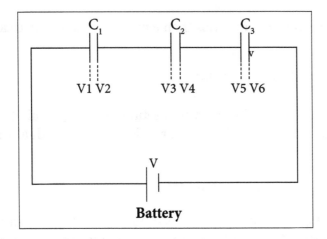

Each plate will have different potential. But the magnitude of charge on the plates is same.

First plate of the C_1 will have potential V_1 which is equal to the voltage of the battery and second plate will have potential less than V_1. Let it be V_2.

Now the first plate of C_2 will have potential equal to V_2 and second plate will have potential less than V_3 let it be V_4.

First plate of C_3 will have potential V_5 ($V_5 = V_4$) and the potential of second plate is less than V_5. Let it be V_6.

But the overall potential difference between the plates is equal to the emf of the battery.

So,

$$V_T = V_1 + V_2 + V_3$$

But we know that,

$$Q = CV$$

$$C = Q/V$$

$$C_{eq} = Q/V_1 + Q/V_2 + Q/V_3 \text{ (As charge is same)}$$

$$1/C_{eq} = (V_1 + V_2 + V_3)/Q$$

$$V_T = Q/C_{eq} = Q/C_1 + Q/C_2 + Q/C_3$$

Hence,

$$1/C_{eq} = 1/C_1 + 1/C_2 + 1/C_3$$

If N capacitors are connected in series then equivalent capacitance can be given as below.

$$1/C_{eq} = 1/C_1 + 1/C_2 + ... + 1/CN$$

Thus when the capacitors are in series connection, the reciprocal of the equivalent capacitance is equal to the sum of the reciprocals of the individual capacitance of the capacitors in the circuit.

Capacitors in Parallel Circuits

There is an advantage of connecting capacitors in parallel than in series. When the capacitors are connected in parallel the total capacitance value is increased. There are some applications where higher capacitance values are required.

Below figure shows the connection of capacitors in parallel. All the positive terminals are connected to one point and negative terminals are connected to another point.

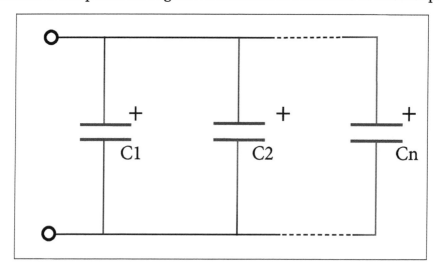

Equivalent Capacitance of the Capacitors in Parallel

All the capacitors which are connected in parallel have the same voltage and is equal to the V_T applied between the input and output terminals of the circuit.

Then, parallel capacitors have a 'common voltage' supply across them .i.e. $V_T = V_1 = V_2$ etc.

The equivalent capacitance, C_{eq} of the circuit where the capacitors are connected in parallel is equal to the sum of all the individual capacitance of the capacitors added together.

This is because the top plate of each capacitor in the circuit is connected to the top plate of adjacent capacitors. In the same way the bottom plate of each capacitor in the circuit is connected to the bottom plate of adjacent capacitors.

Let us see how to calculate the equivalent capacitance of capacitors when connected in parallel. Consider two capacitors connected as shown in the below circuit:

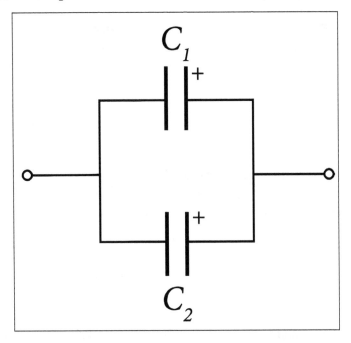

The total charge (Q) across the circuit is divided between the two capacitors, means the charge Q distributes itself between the capacitors connected in parallel. charge Q is equal to the sum of all the individual capacitor charges.

Thus, $Q = Q_1 + Q_2$

Where Q_1, Q_2 are charges at capacitors C_1 and C_2.

We know that,

$$Q = C_{eq} V_T$$

Here,

$$Q = Q_1 + Q_2$$

$$C_{eq} \; V_T = C_1 \times V_1 + C_2 \times V_2$$

Since,

$$V_T = V_1 + V_2 = V$$

$$C_{eq} \; V_T = C_1 \times V + C_2 \times V$$

$$C_{eq} \; V_T = (C_1 + C_2)V$$

If N capacitors are connected in parallel then $C_{eq} = C_1 + C_2 + C_3 + -- C_n$.

Thus equivalent capacitance of the capacitors which are connected in parallel is equal to the sum of the individual capacitance of the capacitors in the circuit.

Here are some applications where capacitors are connected in parallel:

- In some DC supplies for better filtering small capacitors with superior ripple factor are used. These are connected in parallel to increase the capacitance value.

- This can be used in automotive industries in large vehicles like trams for regenerative braking. This application may require large capacitance values than the capacitance usually available in the market.

1.5 Independent and Dependent Sources

An active two-terminal element which supplies energy to the circuit acts as a source of energy. An ideal voltage source is a circuit element which maintains the prescribed voltage across the terminals regardless of the current flowing in those terminals.

An ideal current source is the circuit element that maintains the prescribed current through its terminals regardless of the voltage across those terminals. These circuit elements do not exist as practical devices since they are only idealized models of actual current and voltage sources.

The ideal voltage and current sources is classified as either independent sources or dependent sources. The independent source establishes the current or voltage in a circuit without relying on the currents or voltages in the circuit. Current or voltage is independent of the source.

The dependent source establishes the current or voltage whose value depends on the value of the voltage or current elsewhere in the circuit.

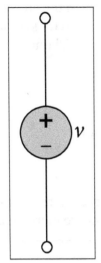

(a) Ideal independent voltage source.

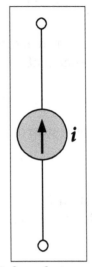

(b) Ideal independent current source.

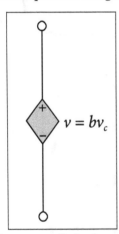

(c) Voltage controlled voltage source.

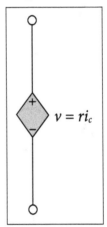

(d) Current controlled voltage source.

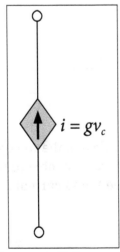

(e) Voltage controlled current source.

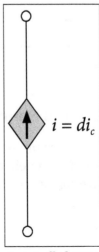

(f) Current controlled current source.

The circuit symbols for ideal independent sources are shown in the above figure (a) and (b). Circle is used to represent the independent source. The circuit symbols for dependent sources are shown in the above figure (c), (d), (e) and (f). Diamond symbol is used to represent a dependent source.

Non-ideal Voltage Sources

Non-ideal voltage source delivers a voltage V and a current i,

$$V = V_s - i.R_s \quad ...(1)$$

The above equation (1) indicates that the voltage delivered by the non-ideal voltage source model decreases as the current out of the voltage source increases.

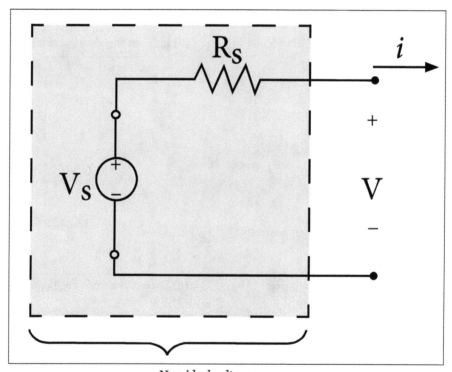

Non-ideal voltage source.

Non-ideal Current Source

An ideal current source provides a specified current regardless of the voltage difference across the device. It can deliver infinite power which is inconsistent with the capabilities of a real current source. The source delivers a voltage V and current i. The output current is given as,

$$i = i_s - \frac{V}{R_s} \quad ...(2)$$

The above equation (2) shows that the current delivered by the source decreases when the delivered voltage increases.

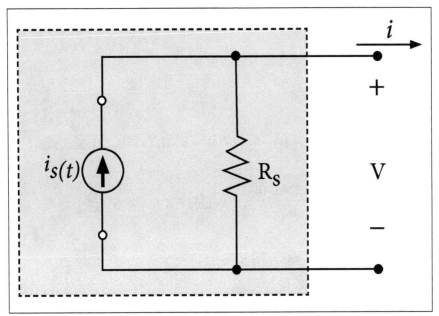

Non-ideal current source model.

1.6   Steady State Solution of DC Circuits

The term DC steady state refers to circuits that have been connected to a DC (voltage or current) source for a very long time where all voltages and currents in the circuits is constant.

$$\frac{di_{L}(t)}{dt}+\frac{R_{1}R_{2}}{L(R_{1}+R_{2})}i_{L}(t)=\frac{R_{2}}{L(R_{1}+R_{2})}\upsilon_{S}(t)$$

$$\frac{L(R_{1}R_{2})}{R_{1}R_{2}}\frac{di_{L}(t)}{dt}+i_{L}(t)=\frac{1}{R_{1}}\upsilon_{S}(t)$$

$$\tau\frac{di_{L}(t)}{dt}+i_{L}(t)=K_{S}\upsilon_{S}(t)$$

Where,

$$\tau=\frac{L(R_{1}+R_{2})}{R_{1}R_{2}}\text{ and }K_{S}=\frac{1}{R_{1}}$$

$$\frac{L(R_1+R_2)}{R_1R_2}\frac{di_L(t)}{dt}+i_L(t)=\frac{1}{R_1}\upsilon_S(t)$$

$$i_L=\frac{1}{R_1}\upsilon_S \text{ as } t\to\infty$$

$$i_L=K_S\upsilon_S$$

$$R_1CL\frac{d^2i_L(t)}{dt^2}+(R_1R_2C+L)\frac{di_L(t)}{dt}+(R_1+R_2)i_L(t)=\upsilon_S(t)$$

$$\frac{R_1CL}{R_1+R_2}\frac{d^2i_L(t)}{dt^2}+\frac{R_1R_2C+L}{R_1+R_2}\frac{di_L(t)}{dt}+i_L(t)=\frac{1}{R_1+R_2}\upsilon_S(t)$$

$$\frac{1}{\omega_n^2}\frac{d^2i_L(t)}{dt^2}+\frac{2\zeta}{\omega_n}\frac{di_L(t)}{dt}+i_L(t)=K_S\,\upsilon_S(t)$$

$$\frac{1}{\omega_n^2}=\frac{R_1CL}{R_1+R_2}\qquad \frac{2\zeta}{\omega_n}=\frac{R_1R_2C+L}{R_1+R_2}\qquad K_S=\frac{1}{R_1+R_2}$$

$$\frac{1}{\omega_n^2}\frac{d^2i_L(t)}{dt^2}+\frac{2\zeta}{\omega_n}=\frac{di_L(t)}{dt}+i_L(t)=Ks\,\upsilon_S(t)$$

$$i_L=Ks\,\upsilon_S \text{ as } t\to\infty$$

$$ic(t)=C\frac{d\,\upsilon_C(t)}{dt}$$

$$ic(t)\to 0 \text{ as } t\to\infty$$

$$\upsilon_L(t)=L\frac{di_L(t)}{dt}$$

$$\upsilon_L(t)\to 0 \text{ as } t\to\infty$$

At DC steady state, all capacitors behave as open circuits and all inductors behave as short circuits.

$$i_C(t)=C\frac{dv_C(t)}{dt}$$

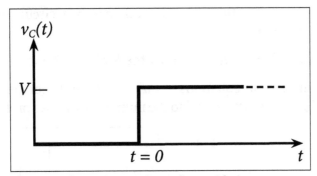

Change in capacitor voltage with time.

The value of an inductor current or a capacitor voltage just prior to the closing (or opening) of the switch is equal to the value just after the switch has been closed (or opened).

$$V_C(0^+) = V_c(0^-)$$

$$i_L(0^+) = i_L(0^-)$$

Where,

0+ is signified as "just after t = 0" and 0- means "just before t = 0"

1.7   Mesh and Nodal Analysis Methods

Mesh Current Method

A mesh is any closed path in a given circuit that does not have any element or branch inside it.

Properties of mesh:

- Every node in the closed path is exactly formed with two branches.
- No other branches are enclosed by the closed path.

Mesh Analysis

This is an alternative structured approach for solving the circuit and it is based on calculating mesh currents. A similar approach to the node situation is used. A set of equations (based on KVL for each mesh) is formed and the equations are solved for unknown values.

Steps for analyzing the mesh current:

- Check whether there is a possibility to transform all current sources in the given circuit to voltage sources.

- Assign the current directions to each mesh in a given circuit and follow the same direction for each mesh.

- Apply KVL to each mesh and simplify the KVL equations.

- Solve the simultaneous equations of various meshes to get the mesh currents and these equations are exactly equal to the number of meshes present in the network.

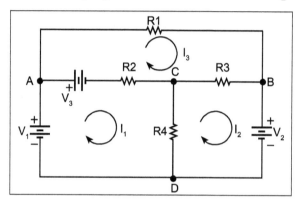

Consider the below DC circuit to apply the mesh current analysis, such that currents in different meshes can be found. In the below figure there are three meshes present as ACDA, CBDC and ABCA but the path ABDA is not a mesh. As a first step, the current through each mesh is assigned with the same direction as shown in the figure.

Secondly, for each mesh we have to apply KVL. By applying KVL around the first loop or mesh we get,

$$V_1 - V_3 - R_2 \left(I_1 - I_3 \right) - R_4 \left(I_1 - I_2 \right) = 0$$

$$V_1 - V_3 = I_1 \left(R_2 + R_4 \right) - I_2 R_4 - I_3 R_2 \qquad \ldots(1)$$

Similarly, by applying KVL around second mesh we get,

$$-V_2 - R_3 \left(I_2 - I_3 \right) - R_4 \left(I_2 - I_1 \right) = 0$$

$$-V_2 = -I_1 R_4 + I_2 \left(R_3 + R_4 \right) - I_3 R_3 \qquad \ldots(2)$$

And by applying KVL around third mesh or loop we get,

$$V_3 - R_1 I_3 - R_3 \left(I_3 - I_2 \right) - R_2 \left(I_3 - I_1 \right) = 0$$

$$V_3 = -I_1 R_2 - I_2 R_3 + I_3 \left(R_1 + R_2 + R_3 \right) \qquad \ldots(3)$$

Therefore, by solving the above three equations we can obtain the mesh currents for each mesh in the given circuit.

Example 1: Let us consider the below example in which we find the voltage across the 12A current source using mesh analysis. In the given circuit all the sources are current sources.

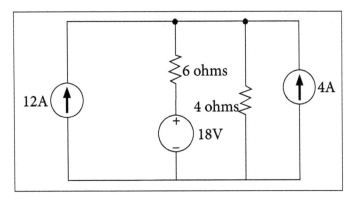

Step 1: In the circuit there is a possibility to change the current source to a voltage source on right hand side source with parallel resistance. The current source is converted into a voltage source by placing the same value of resistor in series with a voltage source and the voltage in that source is determined as,

$$V_S = I_S \ R_S$$

$$= 4 \times 4 = 16 \, \text{Volts}$$

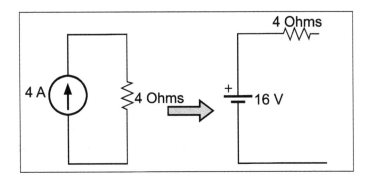

Step 2: Assign the branch currents as I_1 and I_2 to the respective branches or loops and represent the direction of currents as shown below:

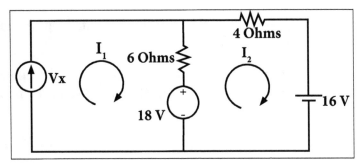

Step 3: Apply the KVL to each mesh in the given circuit.

Mesh 1,

$$V_x - 6 \times (I_1 - I_2) - 18 = 0$$

$$V_x + 6I_2 = 90 \qquad \ldots(1)$$

Mesh 2,

$$18 - 6 \times (I_2 - I_1) - 4 \times I_2 - 16 = 0$$

$$2 - 10 \times I_2 + 6(12) = 0$$

$$I_2 = 74/10$$

$$= 7.4 \, \text{Amps.}$$

Substituting in equation 1 we get,

$$V_x = 90 - 44.4$$

$$= 45.6 \, \text{Volts}$$

Node Voltage Method

In the node voltage method, all the nodes on the circuit are identified. One of them is chosen as the reference voltage (i.e., zero potential) and other node voltages (unknown) are assumed with respect to a reference voltage.

If the circuit has 'n' nodes, then 'n-1' node voltages are unknown. At each of these 'n-1' nodes, we can apply KCL equation. The unknown node voltages become the independent variables of the problem and the solution of node voltages is obtained by solving a set of simultaneous equations.

Nodal Analysis

It determines all the node voltages in the circuit. The voltage at any given node of a circuit is the voltage drop between that node and a reference node.

All the nodes in the circuit are counted and identified. Secondly, nodes at which the voltage is already known are listed. A set of equations based on the node voltages are formed and these equations are solved for unknown quantities.

The set of equations are formed using KCL at each node. The set of simultaneous equations that are produced is then solved. Branch currents can then be found once the node voltages are known.

Steps to Analyse Nodal Analysis Technique:

- Check the possibility to transform voltage sources in the given circuit to the current sources and transform them.

- Identify the nodes present in the given circuit and assign one node as reference node and with respect to this ground or reference node, label other nodes as unknown node voltages.

- Assign the current direction in each branch in the given circuit (it is an arbitrary decision).

- Apply KCL to N-1 nodes and write nodal equations by expressing the branch currents as node assigned voltages.

- Solve the simultaneous equations of nodes to find the node voltages and finally branch currents. The number of node equations is equal to the number of nodes minus one (as one node is referenced).

Consider below DC circuit, in which branch currents are to be determined using the nodal analysis:

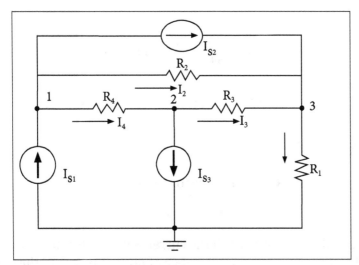

As a first step in nodal analysis, we have to select the reference node which is to be connected to the zero or ground potential as indicated below.

Secondly, we apply Kirchoff's Current Law (KCL) to each node in the circuit except the reference node. By applying KCL at node 1 we get,

$$I_{s1} - I_{s2} - I_4 - I_2 = 0$$

$$I_{s1} - I_{s3} - \{(V_1 - V_2)/R_4\} - \{(V_1 - V_3)/R_2\} = 0$$

$$I_{s1} - I_{s3} = V_1 \{(1/R_2) + (1/R_4) - V_2(1/R_4) - V_3(1/R_2)\}$$

$$I_{s1} - I_{s3} = G_{11}V_1 - G_{12}V_2 - G_{13}V_3 \qquad ...(1)$$

Where, G1i is the sum of total conductance at the first node. (As we know that $1/R = G$)

By applying KCL at node 2 we get,

$$I_4 - I_{s2} - I_3 = 0$$

$$[(V_1 - V_2)/R_4] - I_{s2} - [(V_2 - V_3)/R_3] = 0$$

$$-I_{s2} = -V_1(1/R_4) + V_2[(1/R_3) + (1/R_4) - V_3(1/R_3)]$$

$$-I_{s2} = -G_{21}V_1 - G_{22}V_2 - G_{23}V_3 \qquad ...(2)$$

Applying KCL at node 3 we get,

$$I_{s3} + I_2 + I_3 - I_1 = 0$$

$$I_{s3} + [(V_1 - V_3)/R_2] - [(V_2 - V_3)/R_3] - V_3(1/R_1) = 0$$

$$I_{s3} = -V_1(1/R_2) - V_2(1/R_3) + V_3[(1/R_1) + (1 + R_2) + (1 + R_3)]$$

$$I_{s3} = -G_{31}V_1 - G_{32}V_2 + G_{33}V_3 \qquad ...(3)$$

Likewise, we can write the KCL equations for ith node.

And hence,

$\sum$ Iii is equal to the algebraic sum of all the currents connected at the ith node where i = 1, 2, 3......N and N = n-1 (n is the total number of nodes present in the circuit).

   Gii = The sum of conductance connected to the i th node.

   Gij = the sum of conductance connected between i and j nodes.

By solving the above three equations we get the branch voltages at respective nodes and thereby we can calculate branch currents.

Problems

1. For the circuit shown in the below figure, let us find V_x using the mesh current method.

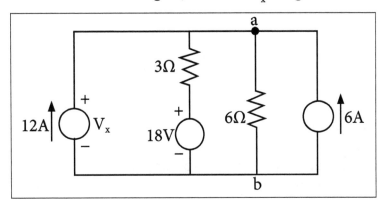

Solution:

Convert the extreme right current source (6A) into a voltage source where the current source magnitude is 6 A and its internal resistance is 6Ω.

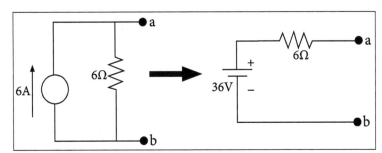

The given circuit is redrawn as shown in the below figure:

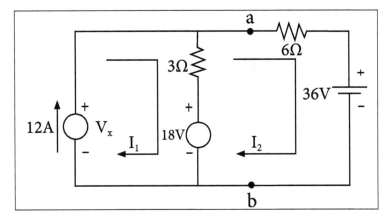

Loop 1: By using KVL, we have,

$$V_x - (I_1 - I_2) \times 3 - 18 = 0 \Rightarrow V_x + 3 I_2 = 54 \qquad \dots(1)$$

Where,

$$I_1 = 12A$$

Loop 2: By using KVL, we have,

$$18 - (I_2 - I_1) \times 3 - I_2 \times 6 - 36 = 0 \Rightarrow 9\,I_2 = 18 \Rightarrow I_2 = 2\,A \quad ...(2)$$

Substituting the equation (2) in equation (1), we get,

$$V_x = 48\text{Volts}$$

2. Let us find the current through branch a-b using mesh analysis.

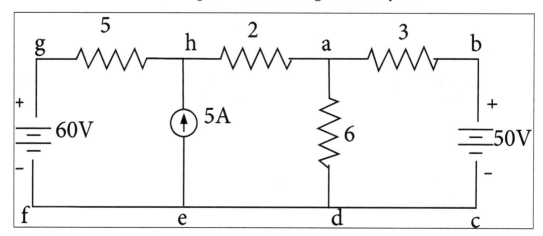

Solution:

Loop: HADEH

$$5I_1 + 2I_2 + 6(I_2 - I_3) = 60$$

$$5I_1 + 8I_2 - 6I_3 = 60 \quad ...(1)$$

Loop: ABCDA

$$3I_3 + 6(I_3 - I_2) = -50$$

$$3I_3 + 6I_3 - 6I_2 = -50$$

$$9I_3 - 6I_2 = -50 \quad ...(2)$$

$$I_2 - I_1 = 5A \quad ...(3)$$

From (1), (2) & (3), we have,

$$D = \begin{vmatrix} -1 & 1 & 0 \\ 5 & 8 & -6 \\ 0 & -6 & 9 \end{vmatrix}$$

$$= -1(72 - 36) - 1(45)$$

$$D = -81$$

$$D_3 = \begin{vmatrix} -1 & 1 & 5 \\ 5 & 8 & 60 \\ 0 & -6 & -50 \end{vmatrix}$$

$$= -1(-400 + 360) - (-250) + 5(-30)$$

$$= 40 + 250 - 150$$

$$D_3 = 140$$

$$I_3 = D_3 / D$$

$$= 140 / -81$$

$$= -1.7283A$$

Current through the branch ab is 1.7283A.

3. Let us find the power in 2Ω and 5Ω resistors from the given circuit by using the nodal analysis method.

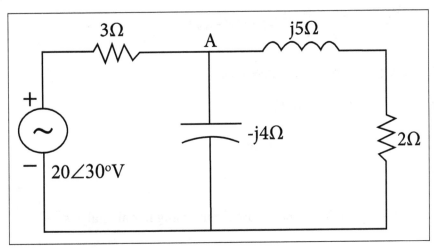

Solution:

By applying nodal analysis for the circuit shown in the above figure, power output of the source and the power in each resistor of the circuit is analyzed as follows.

Step 1: By applying KCL at node A, we have,

$$\frac{V_A - 20 \angle 30°}{3} + \frac{V_A}{-j4} + \frac{V_A}{2+j5} = 0$$

$$V_A \left[\frac{1}{3} - \frac{1}{j4} + \frac{1}{2+j5} \right] - \frac{20\angle 30}{3} = 0$$

$$V_A [0.33 + 0.068 + j0.077] = 6.67 \angle 30°$$

$$V_A [0.398 + j0.077] = 6.67 \angle 30°$$

$$V_A = \frac{6.67 \angle 30°}{[0.398 + j0.078]}$$

$$V_A = \frac{6.67 \angle 30°}{0.41 \angle 11.08°}$$

$$V_A = 16.27 \angle 18.92°$$

Step 2: Current in each resistor is given as,

$$I_{3\Omega} = \frac{V_A - 20\angle 30}{3} = \frac{16.27 \angle 18.92 - 20 \angle 30}{3} = 1.70 \angle -112.2°$$

$$I_{2\Omega} = \frac{V_A}{2+j5} = \frac{16.27 \angle 18.92}{2+j5} = 3.02 \angle -49.27°$$

Step 3: Power in 3 ohm resistor is given as,

$$P_{3\Omega} = I_3^2 * 3 = 1.70^2 * 3 = 8.67 \, W$$

Step 4: Power in 2 ohm resistor is given as,

$$P_{2\Omega} = I_2^2 * 3 = 3.02^2 * 3 = 27.36 \, W$$

4. Let us compute V_1 and V_2 in the below circuit using nodal analysis.

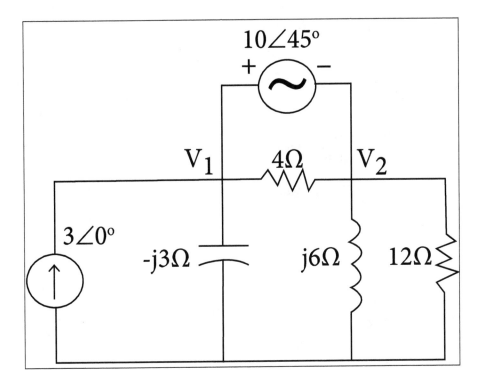

Solution:

Step 1: Voltage difference between V_1 & V_2 is given as,

$$V_1 - V_2 = 10\angle 45°$$

$$V_1 - V_2 = 7.07 + j7.07 \qquad \qquad ...(1)$$

Step 2: By applying KCL at nodes V_1 & V_2, we have,

$$\frac{V_1}{-j3} + \frac{V_2}{J6} + \frac{V_2}{12} = 3\angle 0°$$

$$V_1(j0.33) + V_2\left(\frac{1}{j6} + \frac{1}{12}\right) = 3\angle 0°$$

$$V_1(j0.33) + V_2(0.0833 - 0.1666\,j) = 3 + j0 \qquad \qquad ...(2)$$

Step 3: From equation (1) and (2), we have,

$$\begin{bmatrix} 1 & -1 \\ j0.333 & 0.0833 - j0.1666 \end{bmatrix} \begin{bmatrix} V_1 \\ V_2 \end{bmatrix} = \begin{bmatrix} 7.07 + j7.07 \\ 3 \end{bmatrix}$$

Step 4: To determine V_1 and V_2, we have,

$$V_1 = \frac{\Delta_1}{\Delta} = \frac{\begin{bmatrix} 7.07 + j7.07 & -1 \\ 3 & 0.0833 - j0.1666 \end{bmatrix}}{\begin{bmatrix} 1 & -1 \\ j0.333 & 0.0833 - j0.1666 \end{bmatrix}} = \frac{(0.0833 - j0.1666)*(7.07 + j7.07) + 3}{(0.0833 - j0.1666) + j0.333}$$

$$= \frac{4.7665 - j0.5886}{0.0833 + j0.1664} = 8.636 - j24.322$$

$$V_1 = \frac{\Delta_2}{\Delta} = \frac{\begin{bmatrix} 1 & 7.07 + j7.07 \\ j0.333 & 3 \end{bmatrix}}{\begin{bmatrix} 1 & -1 \\ j0.333 & 0.0833 - j0.1666 \end{bmatrix}} = \frac{3 - (7.07 + j7.07)*(j0.33)}{(0.0833 - j0.1666) + j0.333}$$

$$= \frac{5.354 - j2.354}{0.0833 * j0.1664} = 1.567 - j31.390$$

Node voltages V_1 and V_2 in polar form is given as,

$$V_1 = 8.636 - j24.322 \ (\text{or}) \ 25.80 \angle -70.45°$$

$$V_2 = 1.567 - j31.390 \ (\text{or}) \ 31.429 \angle -87.14°$$

1.7.1 Thevenin's Theorem

Thevenin's theorem for linear electrical networks states that, any combination of voltage sources, current sources and resistors with two terminals is electrically equivalent to the single voltage source V and the single series resistor R. For a single frequency AC systems, it is applied to general impedances but not just for resistors.

The procedure adopted when using the Thevenin's theorem is summarized as follows:

Method 1

Follow these steps in order to find the Thevenin's equivalent circuit, when only the sources of independent type are present:

- Step 1 – Consider the circuit diagram by opening the terminals with respect to which the Thevenin's equivalent circuit is to be found.

- Step 2 – Find Thevenin's voltage V_{Th} across the open terminals of the above circuit.

- Step 3 – Find Thevenin's resistance R_{Th} across the open terminals of the above circuit by eliminating the independent sources present in it.

- Step 4 – Draw the Thevenin's equivalent circuit by connecting a Thevenin's voltage V_{Th} in series with a Thevenin's resistance R_{Th}.

Now, we can find the response in an element that lies to the right side of Thevenin's equivalent circuit.

Method 2

Follow these steps in order to find the Thevenin's equivalent circuit, when the sources of both independent type and dependent type are present:

- Step 1 – Consider the circuit diagram by opening the terminals with respect to which, the Thevenin's equivalent circuit is to be found.

- Step 2 – Find Thevenin's voltage V_{Th} across the open terminals of the above circuit.

- Step 3 – Find the short circuit current ISC by shorting the two opened terminals of the above circuit.

- Step 4 – Find Thevenin's resistance R_{Th} by using the following.

Problems

1. Let us determine the current in the 0.8 Ω resistor using Thevenin's theorem for the below circuit.

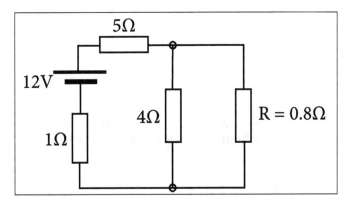

Solution:

By following the above procedure, we have,

(i) 0.8Ω resistor is removed from the above circuit.

Current is given as,

$$I_1 = 12/(1+5+4)$$

$= 12/10 = 1.2A$

(ii) Power dissipation across 4Ω resistor $= 4I_1 = (4)(1.2) = 4.8V.$

Power dissipation across AB i.e., the open-circuit voltage across AB is given as,

$E = 4.8V$

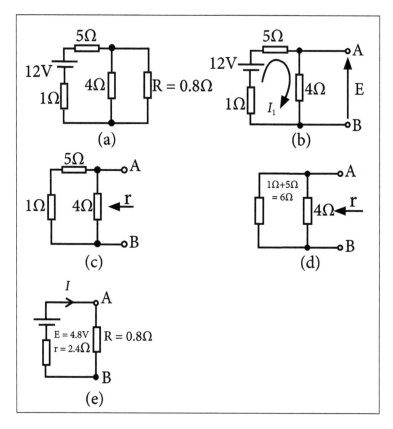

(iii) Removing the source of e.m.f. gives the circuit as shown in the above figure (c). The equivalent circuit of the above figure (c) is shown in the above figure (d) from which resistance r is given as,

$r = (4 \times 6)/(4+6) = 24/10 = 2.4\ \Omega$

(iv) Equivalent Thevenin's circuit is shown in the above figure (e).

From the circuit (e), we have,

Current $I = E/(r + R)$

$= 4.8/(2.4 + 0.8)$

$= 4.8/3.2$ A

$$I = 1.5A = \text{Current in the } 0.8 \ \Omega \text{ resistor}$$

2. Let us find the current flowing through 20 Ω resistor by first finding a Thevenin's equivalent circuit to the left of terminals A and B.

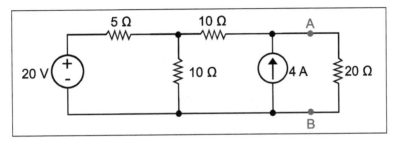

Step 1 – In order to find the Thevenin's equivalent circuit to the left side of terminals A & B, we should remove the 20 Ω resistor from the network by opening the terminals A & B. The modified circuit diagram is shown in the following figure:

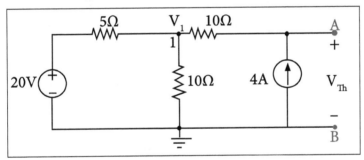

Opening Terminal.

Step 2 – Calculation of Thevenin's voltage V_{Th}.

There is only one principal node except Ground in the above circuit. So, we can use nodal analysis method. The node voltage V_1 and Thevenin's voltage V_{Th} are labelled in the above figure. Here, V_1 is the voltage from node 1 with respect to Ground and V_{Th} is the voltage across 4 A current source.

The nodal equation at node 1 is,

$$\frac{V_1 - 20}{5} + \frac{V_1}{10} - 4 = 0$$

$$\Rightarrow \quad \frac{2V_1 - 40 + V_1 - 40}{10} = 0$$

$$\Rightarrow \quad 3V_1 - 80 = 0$$

$$\Rightarrow \quad V_1 = \frac{80}{3} V$$

The voltage across series branch 10 Ω resistor is,

$$V_{10\Omega} = (-4)(10) = -40\,V$$

There are two meshes in the above circuit:

The KVL equation around second mesh is,

$$V_1 - V_{10\Omega} - V_{Th} = 0$$

Substitute the values of V_1 and $V_{10\Omega}$ in the above equation.

$$\frac{80}{3} - (-40) - V_{Th} = 0$$

$$V_{Th} = 0\frac{80+120}{3} = \frac{200}{3}V$$

Therefore, the Thevenin's voltage is,

$$V_{Th} = \frac{200}{3}V$$

Step 3 – Calculation of Thevenin's resistance R_{Th}.

Short circuit the voltage source and open circuit the current source of the above circuit in order to calculate the Thevenin's resistance R_{Th} across the terminals A & B. The modified circuit diagram is shown in the following figure:

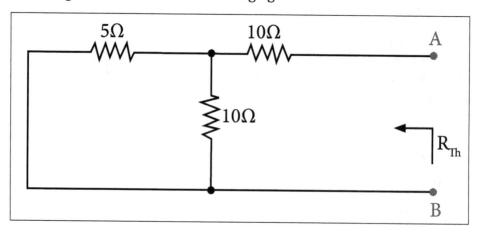

The Thevenin's resistance across terminals A & B will be,

$$R_{Th} = \left(\frac{5\times10}{5+10}\right) + 10 = \frac{10}{3} + 10 = \frac{40}{3}\,\Omega$$

Therefore, the Thevenin's resistance is $R_{Th} = \dfrac{40}{3}\Omega$.

Step 4 – The Thevenin's equivalent circuit is placed to the left of terminals A & B in the given circuit. This circuit diagram is shown in the figure.

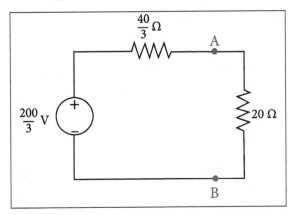

The current flowing through the 20 Ω resistor can be found by substituting the values of V_{Th}, R_{Th} and R in the following equation.

$$I = \frac{V_{Th}}{R_{Th} + R}$$

$$I = \frac{\dfrac{200}{3}}{\dfrac{40}{3} + 20} = \frac{200}{100} = 2A$$

Therefore, the current flowing through the 20 Ω resistor is 2 A.

1.7.2 Norton's Theorem

Any two-terminal linear bilateral DC network is replaced by an equivalent circuit consisting of a current source and a parallel resistor.

Steps in Norton's theorem:

- The portion of the network is removed across which the Norton equivalent circuit is found.

- Terminals of the remaining two-terminal network is marked and RN is found.

- R_N is calculated by setting all sources to zero and the resultant resistance is found between the two marked terminals. Since $R_N = R_{Th}$, the procedure and value obtained using the approach described for Thevenin's theorem will determine the proper value of RN.

- I_N is calculated by returning all the sources to their original position and then the short-circuit current between the marked terminals is found. It is the same amount of current that can be measured by an ammeter which is placed between the marked terminals.

- Norton equivalent circuit is drawn.

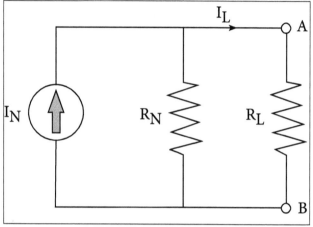

Norton's equivalent circuit.

Current through R_L is given as,

$$I_L = \frac{I_N \cdot R_N}{R_N + R_L}$$

Problems

1. Let us determine the current I flowing in the 4Ω resistance shown in the below figure by using Norton's theorem.

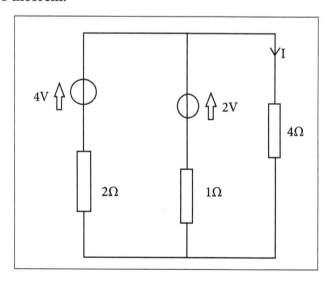

Solution:

4 Ω branch is short-circuited as shown in the below figure:

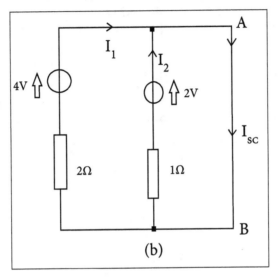

(b)

From the above figure, we have,

$$I_{SC} = I_1 + I_2 = 4A$$

If the sources of e.m.f. are removed, the resistance 'looking-in' at a break made between A and B is given as,

$$r = 2 \times 1/2 + 1 = 2/3 \, \Omega$$

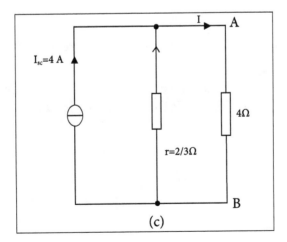

(c)

From the Norton equivalent network shown in the above figure (c), the current in the 4 Ω resistance is given as,

$$I = \left[(2/3)/((2/3) + 4)\right](4) = 0.571A$$

2. Let us find the current flowing through 20 Ω resistor by first finding a Norton's equivalent circuit to the left of terminals A and B.

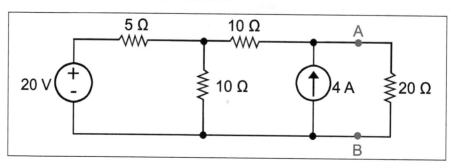

Solution:

Step 1 – Consider the below figure.

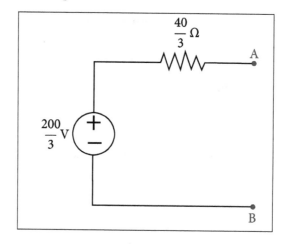

Here, Thevenin's voltage, $V_{Th} = \dfrac{200}{3} V$ and Thevenin's resistance, $R_{Th} = \dfrac{40}{3} \Omega$

Step 2 – Apply source transformation technique to the above Thevenin's equivalent circuit. Substitute the values of VTh and RTh in the following formula of Norton's current.

$$I_N = \frac{V_{Th}}{R_{Th}}$$

$$I_N = \frac{\dfrac{200}{3}}{\dfrac{40}{3}} = 5A$$

Therefore, Norton's current I_N is 5 A.

We know that Norton's resistance, R_N is same as that of Thevenin's resistance R_{Th}.

$$R_N = \frac{40}{3}\Omega$$

The Norton's equivalent circuit corresponding to the above Thevenin's equivalent circuit is shown in the following figure:

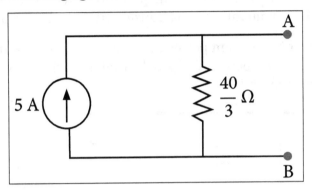

Now, place the Norton's equivalent circuit to the left of the terminals A & B of the given circuit.

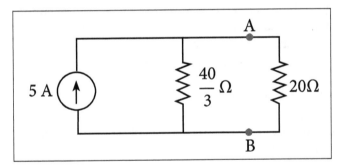

By using current division principle, the current flowing through the 20 Ω resistor will be,

$$I_{20\Omega} = 5\left(\frac{\dfrac{40}{3}}{\dfrac{40}{3}+20}\right)$$

$$I_{20\Omega} = 5\left(\frac{40}{100}\right) = 2A$$

Therefore, the current flowing through the 20 Ω resistors is 2 A.

1.7.3 Maximum Power Transfer Theorem

It is used in designing electrical network analysis and electrical circuit. It was invented by a German engineer Moritz von Jacobi in the year 1840. He invented the theorem to maximize the output of the battery to a motored boat which travels in the river Neva. Hence, the theorem is referred as Jacobi's law.

It deals with the transfer of power to the load on a circuit with a network of various sources or components on it. The maximum power transfer theorem defines the condition under which the maximum power is transferred to the load in a circuit.

Under the condition of maximum power transfer, we only deal with the power transferred to the load and the power dissipated in internal circuits or resistance of the source is not determined.

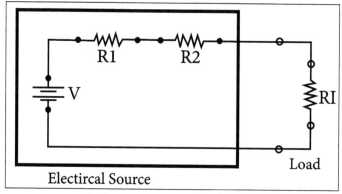

Maximum power transfer theorem.

According to maximum power transfer theorem, maximum power is yielded to the load R_L where R_L is equal to the internal resistance of the circuit (R_1+R_2). The maximum power transfer theorem is true in case of any kind of circuit which may be linear, non-linear, active, DC or AC.

In DC circuits, load resistance is matched with the internal resistance of the source where both resistances are equal and in AC, the load impedance is matched with the internal impedance of the circuit or source where the load impedance is constant.

For example, load impedance will be R1 - jX.

Proof of maximum power transfer theorem:

   V = emf supplied to the load.

   R_L = Load resistance.

   R_i = Internal resistance of the source.

   I = Current flowing through the load, internal resistance and the source of the circuit.

   P_L = Power transferred to the load.

P_i = Power dissipated at the internal resistance.

Power transferred to the load $= P_L = I^2 R_L$

$$P_L = \left(\frac{V}{R_i + R_L} \right)^2 \times R_L = \frac{V^2}{\dfrac{R_i^2}{R_L} + 2R_i + R_L}$$

Using the theorems of differential calculus, if R_L is variable, then the maximum value of P_L is calculated and P_L is differentiated with respect to R_L and it is equated with zero. Thus, under maximum power transfer to load condition, we have,

$$\frac{d}{dR_L} P_L = \frac{d}{dR_1} \frac{V^2}{\dfrac{R_i^2}{R_L} + 2R_i + R_L} = 0$$

$$-\frac{R_i^2}{R_L^2} + 1 = 0$$

$$R_i = R_L$$

$Z_i = R_i + X_i = $ Internal impedance of reactive circuits

$Z_L = R_L + X_L = $ Load impedance

Under the condition of maximum power transfer to load, we have,

$$Z_i = Z_1$$

$$R_i = R_L$$

$$X_i = -X_L$$

Efficiency of Power Transfer

It is the efficiency of any source or circuit in transferring its power to the load or it is the ratio of power transferred to the load over total power transferred by the source. It is denoted as η.

$$\eta = P_L / P_T$$

Where,

P_L = Power transferred to the load.

P_T = Total power transferred by the source.

$P_L = I^2 R_L$

$P_T = P_L + P_I = I^2 R_L + I^2 R_I$

Where,

I = Total current flowing through the circuit or Thevenin's equivalent of the circuit

P_I = Power dissipated in internal circuits of the source,

$$\eta = \frac{P_L}{P_T} = \frac{I^2 R_L}{I^2 R_L + I^2 R_I} = \frac{R_L}{R_L + R_I} = \frac{1}{1 + \dfrac{R_I}{R_L}}$$

Under the condition of maximum power transfer,

$R_L = R_I$

Efficiency is 0.5 under the condition of maximum power transfer.

Overall efficiency decreases if the value of R_L is kept very low and it increases up to one when the value of R_L is increased to infinity as shown on the graph below and the power transferred to load becomes minimum when R_L is kept very low and very high.

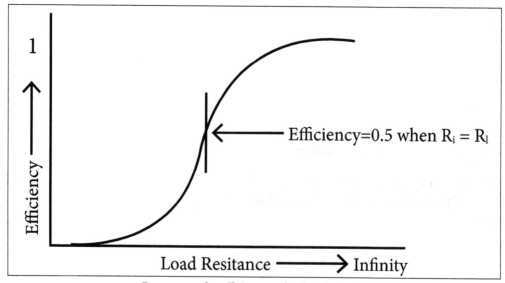

Power transfer efficiency vs load resistance.

Maximum power transfer to load is obtained when $R_L = R_I$ as shown on the below graph:

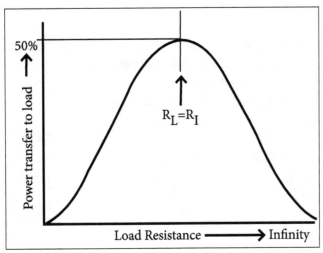

Power transfer to load vs load resistance.

Steps for Solving Network using Maximum Power Transfer Theorem

Following steps are used to solve the problem by Maximum Power Transfer theorem:

- Step 1: Remove the load resistance of the circuit.

- Step 2: Find the Thevenin's resistance (R_{TH}) of the source network looking through the open circuited load terminals.

- Step 3: As per the maximum power transfer theorem, this RTH is the load resistance of the network, i.e., $R_L = R_{TH}$ that allows maximum power transfer.

- Step 4: Maximum Power Transfer is calculated by the equation shown below,

$$P_{max} = \frac{V_{Th}^2}{4R_{Th}}$$

Applications of Maximum Power Transfer Theorem

Maximum power transfer theorem has a wide range of usage on real life situation. This theorem maximizes the power output to a load from any circuit. They are used to design circuits where the maximum output performance is desired. For example, if we match the amplifier with the loudspeaker, it will yield the maximum power to the speaker and it produces maximum sound.

Transformer coupling yields maximum power to the load when the matching of load and source impedance is not possible. The application of maximum power theorem is done only under the conditions when the maximum performance is desired over the overall efficiency of the circuit because the efficiency of the circuit under maximum power transfer condition is only 0.5.

Problems

1. If 8Ω loudspeaker is connected to an amplifier with an output impedance of 1000Ω, let us calculate the turns ratio of the matching transformer to provide maximum power transfer of the audio signal. Let us assume that the source impedance of an amplifier is Z_1, the load impedance is Z_2 where the turns ratio is N.

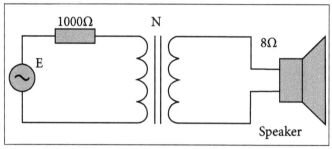

Matching circuit of transformer impedance.

Solution:

Given:

   8Ω Loudspeaker

   Output impedance - 1000Ω

   Source impedance of an amplifier - Z_1

   Load impedance - Z_2

Transformer turns ratio is given as,

$$Z_1 = N^2 Z_2 \quad \therefore \quad N = \sqrt{\frac{Z_1}{Z_2}}$$

Therefore,

$$N = \sqrt{\frac{Z_1}{Z_2}} = \sqrt{\frac{1000}{8}} = 11.2 : 1$$

Small transformers used in low power audio amplifiers are ideal and any losses can be ignored.

2. Let us determine the value of R_L for the given network where the power is maximum. Let us also determine the maximum power through the load-resistance R_L by using maximum power transfer theorem.

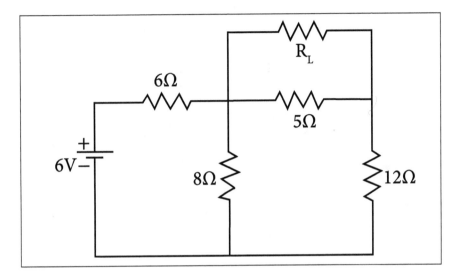

Solution:

Let us determine the value of unknown resistance (R_L). When power is maximum through load, resistance is equal to the equivalent resistance between the two ends of load-resistance. Let us determine the equivalent resistance to find the load resistance R_L.

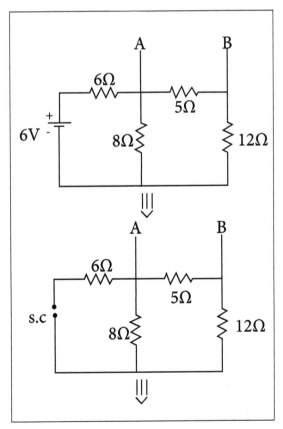

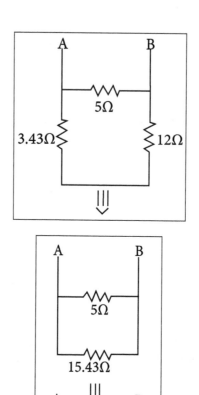

$$R_{AB} / R_L = 3.77\Omega$$

To find the maximum power through load-resistance, let us find-out $V_{o.c}$.

$V_{o.c}$ is known as voltage between open circuits.

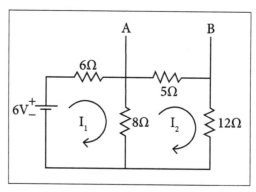

Applying KVL in 1st loop, we have,

$$6 - 6I_1 - 8I_1 + 8I_2 = 0$$

$$-14I_1 + 8I_2 = -6 \qquad \ldots(1)$$

Applying KVL in 2nd loop, we get,

$$-8I_2 - 5I_2 - 12I_2 + 8I_1 = 0$$

$$8I_1 - 25I_2 = 0 \qquad \ldots(2)$$

On solving equations (1) and (2), we get,

$$I_1 = 0.524 \text{ A}$$

$$I_2 = 0.167 \text{ A}$$

$$V_A - 5I_2 - V_B = 0$$

$$V_{o.c} / V_{AB} = 5I_2$$

$$= 5 \times 0.167$$

$$= 0.835V$$

Maximum power through R_L is given as,

$$P_{max} = \frac{V_{o.c}^2}{4R_L}$$

$$P_{max} = \frac{0.835^2}{4 \times 3.77}$$

$$P_{max} = 0.046 \text{ Watt}$$

3. Let us find the maximum power that can be delivered to the load resistor R_L of the circuit shown in the following figure.

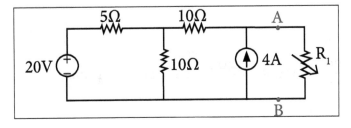

Solution:

Step 1: Consider the following figure.

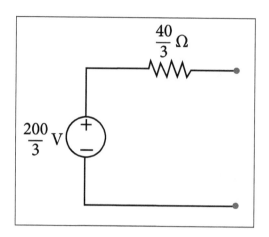

Here, Thevenin's voltage $V_{Th} = \dfrac{200}{3}V$ and Thevenin's resistance $R_{Th} = \dfrac{40}{3}\Omega$

Step 2 – Replace the part of the circuit, which is left side of terminals A & B of the given circuit with the above Thevenin's equivalent circuit. The resultant circuit diagram is shown in the following figure.

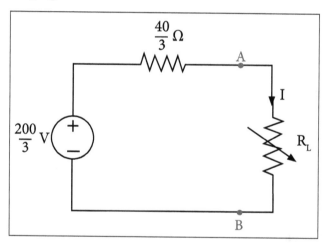

Step 3 – we can find the maximum power that will be delivered to the load resistor, R_L by using the following formula.

$$P_{L,Max} = \frac{V_{Th}{}^2}{4R_{Th}}$$

Substitute $V_{Th} = \dfrac{200}{3}V$ and $R_{Th} = \dfrac{40}{3}\Omega$ in the above formula,

$$P_{L,Max} = \frac{\left(\dfrac{200}{3}\right)^2}{4\left(\dfrac{40}{3}\right)}$$

$$P_{L,Max} = \frac{250}{3} W$$

Therefore, the maximum power that will be delivered to the load resistor RL of the given circuit is $\frac{250}{3}$ W.

4. Consider the below circuit for which we are going to determine the value of load resistance, R_L for which maximum power will transfer from source to load.

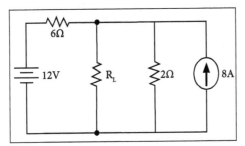

Now, the given circuit can be further simplified by converting the current source into equivalent voltage source as follows:

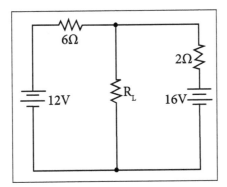

We need to find the Thevenin's equivalent voltage V_{th} and Thevenin's equivalent resistance R_{th} across the load terminals in order to get the condition for maximum power transfer. By disconnecting the load resistance, the open-circuit voltage across the load terminals can be calculated as;

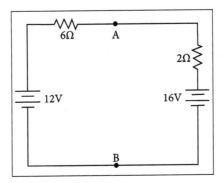

By applying Kirchhoff's voltage law, we get,

$$12 - 6I - 2I - 16 = 0$$

$$-8I = 4$$

$$I = -0.5 \text{ A}$$

The open-circuit voltage across the terminals A and B, $V_{AB} = 16 - 2 \times 0.5 = 15$ V

Thevenin's equivalent resistance across the terminals A and B is obtained by short-circuiting the voltage sources as shown in the below figure:

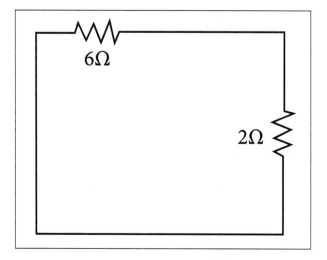

$$R_{eq} = (6 \times 2)/(6+2)$$

$$= 1.5 \ \Omega$$

So the maximum power will transferred to the load when R_L = 1.5 ohm.

Current through the circuit, $I = 15/(1.5 + 1.5) = 5$ A

Therefore, the maximum power $= 5^2 \times 1.5 = 37.5$ W

1.7.4 Linearity and Superposition Theorem

Superposition theorem is one of the electrical network analysis theorem which helps us to solve the linear circuit with more than one current or voltage source easily.

In a linear circuit with several sources, the current and voltage responses in any branch is the algebric sum of the voltage and current responses where each of the source acts independently where all the other sources are replaced by their internal impedance.

Suppose an electrical circuit having several branches or loads and also several sources with some being current source and some being voltage source, then superposition theorem suggests the following.

If we find the branch responses on a branch due to only one of the source by ignoring the effect of all the other sources or replacing all the other sources by their corresponding internal impedance and the process is repeated for every source on the circuit, then the combined responses (voltage drop and current through it) on a branch due to all the sources combined is the algebric sum of the responses on the branches due to each individual sources.

Steps to be followed in the superposition theorem:

Step 1: Any one of the energy source is selected and all the other energy sources are replaced by their internal series resistances.

Step 2: With only one energy source, let us calculate the current through the required branch by using series parallel along with current division rule or by using mesh analysis.

Step 3: Repeat the steps 1 and 2 for each source individually.

Step 4: The currents obtained due to the individual source are added to obtain the combined effect of all the sources.

Problems

1. Let us determine the current I_{ab} flowing through the 3 Ω resistor as shown in the below figure using the superposition theorem.

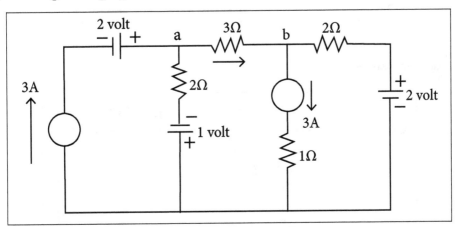

Solution:

Assume that the current source 3A is acting alone in the circuit where the other sources are replaced by their internal resistance.

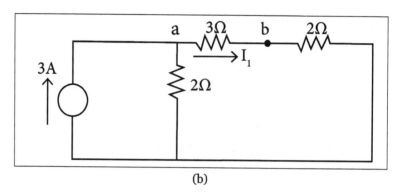

(b)

Current flowing through 3 Ω resistor is given as,

$$I_{1(\text{due to 3A current source})}=3\times\frac{2}{7}=\frac{6}{7}A(a\text{ to }b) \qquad ...(1)$$

Current flowing through 3Ω resistor due to 2V source is obtained from the below figure(c):

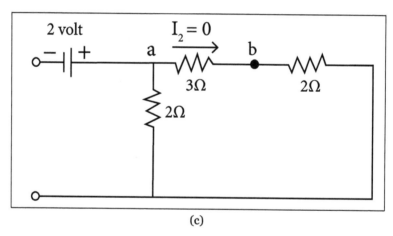

(c)

$$I_{2(\text{due to 2V voltage source})}=0A$$

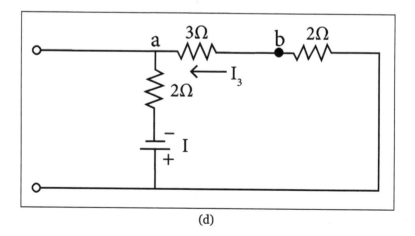

(d)

Current through 3Ω resistor due to 1V voltage source is given as,

$$I_{3(\text{due to 1V voltage source})} = 1/7 \text{ A } (\text{b to a})$$

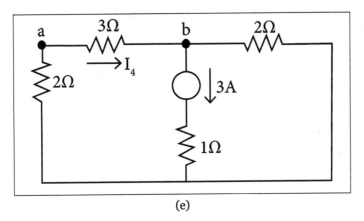

(e)

From the above figure (e), current through 3Ω resistor due to 3A current source is given as,

$$I_{4(\text{due to 3A current source})} = 3 \times \frac{2}{7} = \frac{6}{7} \text{ A } (\text{a to b})$$

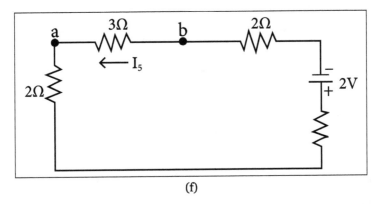

(f)

From the above figure (f), current through 3 Ω resistor due to 2V voltage source is given as,

$$I_{5(\text{due to 2V voltage source})} = \frac{2}{7} \text{ A } (\text{b to a}) \qquad ...(2)$$

Resultant current I_{ab} flowing through 3Ω resistor due to the combination of all sources is obtained by the following expression.

$$I_{ab} = I_{1(\text{due to 3A current source})} + I_{2(\text{due to 2V current source})} + I_{3(\text{due to 1V current source})}$$
$$+ I_{4(\text{due to 3A current source})} + I_{5(\text{due to 3A current source})}$$

$$= \frac{6}{7} + 0 - \frac{1}{7} + \frac{6}{7} - \frac{2}{7} = \frac{9}{7} = 1.285 \, (\text{a to b})$$

2. Using the superposition theorem, let us determine the voltage drop and current across the 3.3KΩ as shown in the below figure:

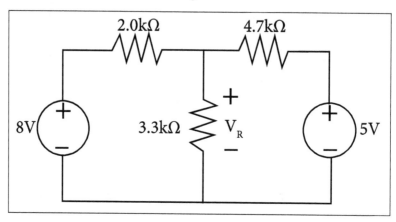

Solution:

Step 1: 8V power supply is removed from the original circuit where the new circuit is obtained as in the following figure and voltage across the resistor is measured.

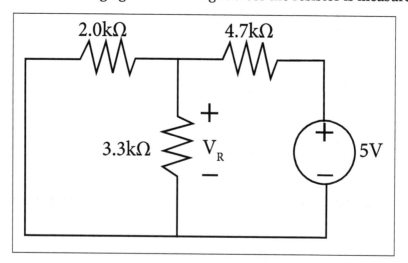

Since 3.3KΩ and 2KΩ are parallel, the resultant resistance will be 1.245KΩ.

Using voltage divider rule, voltage across 1.245KΩ is given as,

$$V_1 = \left[1.245 / (1.245 + 4.7) \right] * 5 = 1.047 \text{V}$$

Step 2: 5V power supply is removed from the original circuit where the new circuit is shown in the below figure and voltage across the resistor is measured.

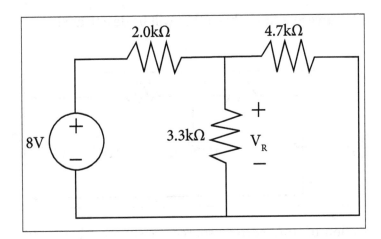

Since 3.3KΩ and 4.7KΩ are parallel, the resultant resistance will be 1.938KΩ.

Using voltage divider rule, the voltage across 1.938KΩ is given as,

$$V_2 = \left[1.938/(1.938+2)\right]*8 = 3.9377V$$

Voltage drop across 3.3KΩ resistor is given as,

$$V_1 + V_2 = 1.047 + 3.9377 = 4.9847V$$

3. Let us find the current through 10 Ω resistance in the given network by using super-position theorem.

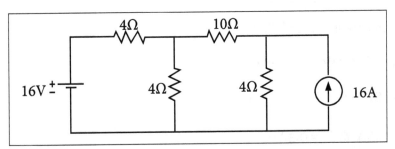

Solution:

Step 1: Let us find current through 10Ω resistance by using superposition theorem.

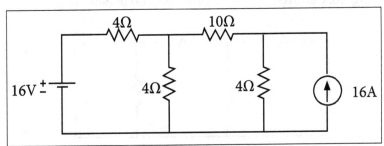

If we activate 16V source at a time, then other will be deactivated.

Step 2:

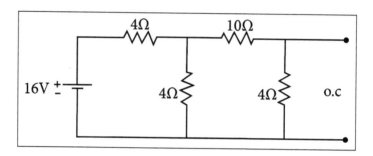

After the deactivation of '16A' current source, mesh analysis is used to find current through 10Ω resistance when '16V' voltage source is active. We can use nodal analysis or ohm's law with current division rule.

Step 3:

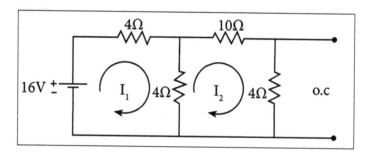

Mesh equations are given as follows.

$$-8I_1 + 4I_2 = -16 \quad \ldots(1)$$

$$4I_1 - 18I_2 = 0 \quad \ldots(2)$$

$$I_1 = 2.25A \ \& \ I_2 = 0.5A$$

$$I_{10\Omega} = I_2 = 0.5 \ A$$

Step 4: Activating '16A' source at a time, other will be deactivated.

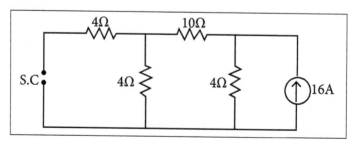

Step 5: After deactivation of '16V' voltage source, by applying current division rule, we can easily find the value of current in 10Ω resistance.

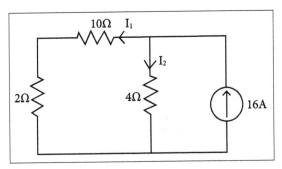

By current division rule, we have,

$$I_1 = 4A$$

$$I_2 = 12A$$

$$I_{10\Omega} = I_1 = 4A$$

Step 6: In the final circuit, all the sources are activated. Direction of current when single source is active and other is deactivated through 10Ω resistance will be shown with their value as shown in the below figure:

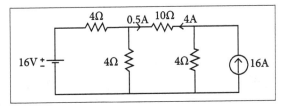

Step 7: Current through 10Ω resistance is 3.5A.

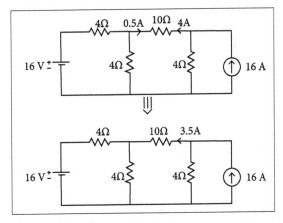

Current flowing through 10 Ω = 3.5 A.

2

AC Circuits

2.1   AC Circuits: Waveforms and RMS Value

AC Circuits

Alternating current (A.C) is the current that flows in one direction for a brief time which reverses and flows in opposite direction for a similar time. The source for alternating current is called as AC generator or alternator.

Fundamentals of AC

An alternating (ac) quantity (voltage, current or power) is defined as the one which changes its value as well as direction (polarity) with respect to time. All our appliances such as TV, refrigerators, washing machines, air conditioners, fans etc., operate on the alternating voltage (ac voltage).

Waveforms and RMS Value

An alternating quantity changes continuously in magnitude and alternates in direction at regular intervals of time.

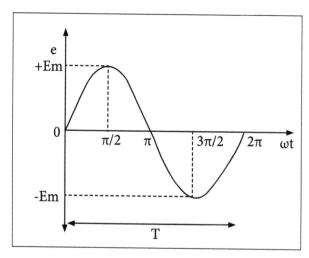

Amplitude: It is the maximum value attained by an alternating quantity. It is also called as maximum or peak value.

Instantaneous Value: It is the value of the quantity at any instant.

Time Period (T): It is the time taken in seconds to complete one cycle of an alternating quantity.

Frequency (f): It is the number of cycles that occur in one second. The unit for frequency is Hz or cycles/sec. The relationship between frequency and time period can be derived as follows,

> Time taken to complete f cycles = 1 second
>
> Time taken to complete 1 cycle = 1/f second
>
> $T = 1/f$

Advantages of AC system over DC system:

- AC motors are cheaper and simpler in construction than DC motors.
- Switchgear for AC system is simpler than DC system.
- AC voltages can be efficiently stepped up/down using transformer.

Energy storage elements:

- Capacitor (C).
- Inductor (L).

Capacitors and inductors are energy storing elements. Capacitors stores electrical energy while inductors store magnetic energy.

Capacitor: A capacitor is a passive element which stores energy in its electric field.

Inductor: An inductor is a passive electrical device which stores energy in a magnetic field by combining the effects of many loops of an electric current. The inductance is measured in Henry (H).

2.1.1 Power and Power Factor

Power in AC Circuits

50Hz or 60Hz AC electric power constitutes the most common form of electric power distribution where the phasor notation analyzes the power absorbed by both resistive and complex loads.

Three forms of powers in AC circuits are:

- Active power or True power or Real power.

- Reactive power.

- Apparent power.

Active Power

The actual amount of power being dissipated or performs the useful work in the circuit is called as active or true or real power. It is measured in watts, practically measured in KW (kilowatts) and MW (megawatts) in power systems.

It is denoted by the letter P (capital) and it is equal to the average value of $P = VI \cos \varphi$. It is the desired outcome of an electrical system which drives the circuit or load.

$$P = VI \cos \varphi$$

Reactive Power

The average value of the second term in the above derived expression is zero, so the power contributed by this term is zero. The component, which is proportional to VI sin ϕ is called as reactive power, represented by the letter Q.

Even though it is a power, but not measured in watts as it is a non-active power and hence, it is measured in Volt-Amperes- Reactive (VAR). The value of this reactive power can be negative or positive depends on the load power factor.

This is because inductive load consumes the reactive power while capacitive load generates the reactive power.

$$Q = VI \sin \varphi$$

Apparent Power

The complex combination of true or active power and reactive power is called apparent power. Without reference to any phase angle, the product of voltage and current gives the apparent power. The apparent power is useful for rating the power equipment.

It can also be expressed as the square of the current multiplied by the circuit's impedance. It is denoted by the letter S and measured in Volt-Amperes (VA), practical units include KVA (Kilo volt-amperes) and MVA (mega volt-amperes).

Power Factor

It is the cosine of the phase angle between current and voltage. i.e.,

$$\text{Power factor} = \cos \theta$$

Power factor is the ratio of true power to the apparent power.

$$\text{Power Factor} = \frac{\text{True power}}{\text{Apparent power}}$$

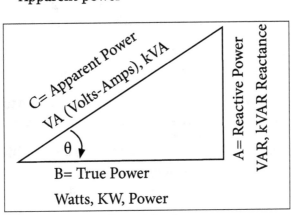

Power triangle.

In pure resistive circuit, current is in phase with circuit voltage (i.e.,) $\phi = 0$.

Therefore, power factor of the resistive circuit is given as,

$\cos \phi = 1$

In pure inductive or capacitive circuit, current is 90° out of phase with circuit voltage (i.e.,) $\phi = 90°$.

Power factor of the inductive circuit is given as,

$\cos \phi = 0$

For circuits having resistance - inductance, resistance - capacitance or resistance - inductance - capacitance, the power factor lies between 0 and 1. It is noted that the value of power factor will never be more than one.

The word lagging or leading is attached with the numerical value of power factor to signify whether the current lags or leads the voltage. In inductive circuits, current always lags behind the voltage and their power factors are mentioned as lagging power factor. For capacitive circuits, the power factor is mentioned as leading power factor since current always leads the voltage vector.

Importance of Power Factor

The power factor of an AC circuit plays an important role in power system.

The power of an AC circuit is given as,

$$P = VI \cos \phi = \text{or } I = \frac{P}{V \cos \phi}$$

Disadvantage of an AC Circuit

Greater Conductor Size: At low power factor, the conductors have to carry more current for the same power and they require large area of cross-section.

Poor Efficiency: At low power factors, the conductors have to carry larger current which increases copper losses (I2R) and results in poor efficiency.

Larger Voltage Drop: At low power factors, the conductors have to carry large currents that increases voltage drop (IR) in the system and results in poor regulation.

Larger kVA Rating of Equipment: kVA rating of electrical machines and equipment's connected in the power system (alternators, transformers and switchgears) will be more at low power factors since it is inversely proportional to power factor (i.e., kVA = kW/cos ϕ).

To improve the power factor of an AC circuit, a capacitor is connected across the circuit (i.e.,) parallel to the circuit.

Problems

1. If an AC power supply of 100V, 50Hz is connected across a load of impedance, 20 + j15 Ohms. Then let us calculate the current flowing through the circuit, active power, apparent power, reactive power and power factor.

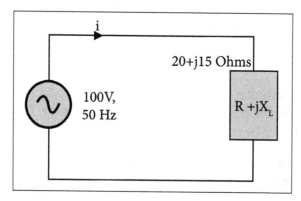

Given that,

$$Z = R + jXL = 20 + j\ 15\ \Omega$$

Converting the impedance to polar form, we get

$$Z = 25\ \angle 36.87\ \Omega$$

Current flowing through the circuit,

$$I = V/Z = 100 \angle 00 / 25\ \angle 36.87$$

$$I = 4 \angle -36.87$$

Active power, $P = I2R = 42 \times 20 = 320$ watts

or,

$$P = VI \cos \varphi = 100 \times 4 \times \cos(36.87) = 320.04 \approx 320 \text{ W}$$

Apparent power, $S = VI = 100 \times 4 = 400$ VA

Reactive power, $Q = \sqrt{(S^2 - P^2)}$

$$= \sqrt{(4002 - 3202)} = 240 \text{ VAr}$$

Power factor, $PF = \cos \varphi = \cos 36.87 = 0.80$ lagging.

2. Two loads of 10KW each are operating at a power factor 0.8 lagging (each). Let us find their combined power factor.

Solution:

In the above question, it is not specified whether the loads are connected in series or parallel, our answer does not depend whether it is connected in series or parallel because in both the case the answer will be same.

$$\text{Power Factor} = \frac{\text{True Power}}{\text{Apparent Power}}$$

$$\text{Apparent Power} = \frac{\text{True Power}}{\text{Power Factor}}$$

For Load 1,

$$\text{Apparent Power} = \frac{10\,\text{KW}}{0.8}$$

$$\text{Apparent Power} = 12.5\,\text{KVA}$$

For Load 2,

$$\text{Apparent Power} = 12.5\,\text{KVA}$$

Combined Power Factor,

$$\text{Combined Power Factor} = \frac{\text{Total True Power}}{\text{Total Apparent Power}}$$

$$\text{Combined Power Factor} = \frac{10+10}{12.5+12.5}$$

Combined Power Factor $=0.8\,\text{lag}$

3. A wound coil that has an inductance of 180mH and a resistance of 35Ω is connected to a 100V 50Hz supply. Let us calculate: a) The impedance of the coil, b) The current, c) The power factor, and d) The apparent power consumed. Let us also draw the resulting power triangle for the above coil.

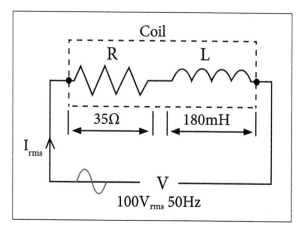

Solution:

Data given:

R = 35Ω, L = 180mH, V = 100V and f = 50Hz.

1. Impedance (Z) of the coil:

$$R=35\Omega$$

$$X_L = 2\pi fL = 2\pi \times 50 \times 0.18 = 56.6\Omega$$

$$Z=\sqrt{R^2 + X_L^2} = \sqrt{35^2 + 56.6^2} = 66.5\Omega$$

2. Current (I) consumed by the coil:

$$V=I\times Z$$

$\therefore \qquad I = \dfrac{V}{Z} = \dfrac{100}{66.5} = 1.5\,A\,(\text{rms})$

3. The power factor and phase angle, θ:

$$\cos\theta = \frac{R}{S}, \text{ or } \sin\theta = \frac{X_L}{Z}, \text{ or } \tan\theta = \frac{X_L}{R}$$

$$\therefore \quad \cos\theta = \frac{R}{Z} = \frac{35}{66.5} = 0.5263$$

$$\cos\ (0.5263) = 58.2°(\text{lagging})$$

4. Apparent power (S) consumed by the coil:

$$P = V \times I \cos\theta = 100 \times 1.5 \times \cos(58.2°) = 79\,W$$

$$Q = V \times I \sin\theta = 100 \times 1.5 \times \sin(58.2°) = 127.5\,VAr$$

$$S = V \times I = 100 \times 1.5 = 150\,VA$$

or,

$$S^2 = P^2 + Q^2$$

$$\therefore \quad S = \sqrt{P^2 + Q^2} = \sqrt{79^2 + 127.5^2} = 150\ VA$$

5. Power triangle for the coil:

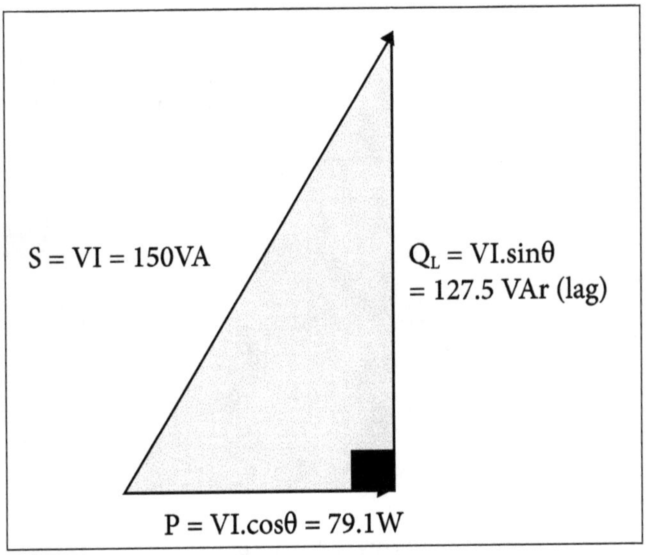

S = VI = 150VA

$Q_L = VI.\sin\theta$ = 127.5 VAr (lag)

P = VI.cosθ = 79.1W

As the power triangle relationships of this simple example demonstrates, at 0.5263 or 52.63% power factor, the coil requires 150 VA of power to produce 79 Watts of useful work. In other words, at 52.63% power factor, the coil takes about 89% more current to do the same work, which is a lot of wasted current.

Adding a power factor correction capacitor (for this example a 32.3uF) across the coil, in order to increase the power factor to over 0.95, or 95%, would greatly reduce the reactive power consumed by the coil as these capacitors act as reactive current generators, thus reducing the total amount of current consumed.

2.2   Single Phase and Three Phase Balanced Circuits

Single Phase Power Systems

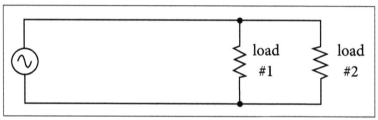

Single phase power systems.

The above figure is a very simple AC circuit. If the power dissipation of load resistor were substantial, we might call this as "power circuit" or "power system" instead of regarding it as just a regular circuit. Size and cost of wiring is necessary to deliver power from the AC source to the load.

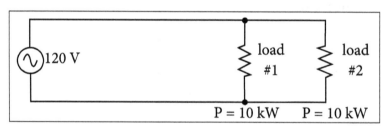

$$l=\frac{P}{E}$$

$$l=\frac{10\,kW}{120\,V}$$

$$l=83.33\,A$$

$$l_{total}=l_{load\#1}+l_{load\#2} \qquad P_{total}=(10\,kW)+(10\,kW)$$

$$l_{total}=(83.33\,A)+(83.33\,A) \qquad P_{total}=20\,kW$$

$$l_{total}=166.67\,A$$

Three Phase Balanced Circuits

The voltages in the three-phase power system are produced by a synchronous genera-tor. In a balanced system, each of the three instantaneous voltages has equal amplitude but they are separated from the other voltages by a phase angle of 1200. The three volt-ages are labeled as a, b and c. The common reference point for the three phase voltages is designated as the neutral connection which is labeled as n.

We may define either a positive phase sequence (abc) or a negative phase an, bn, cn sequence (acb) as shown in the below figure. The three sources V_{an}, V_{bn} and V_{cn} are designated as the line-to-neutral voltages in the three-phase system.

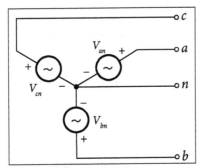

Balanced system.

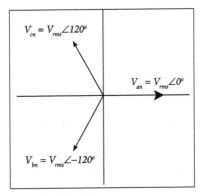

Positive phase sequence.

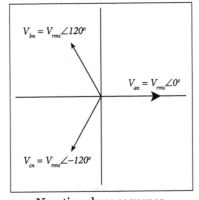

Negative phase sequence.

Line-to-line Voltages

The alternative way of defining the voltages in a balanced three-phase system is to define the voltage differences between the phases. These voltages are designated as line-to-line voltages which can be expressed in terms of line-to-neutral voltages by applying Kirchhoff's voltage law to the generator circuit as follows:

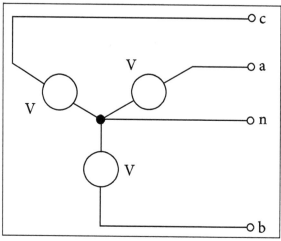

Line to line voltages.

$$V_{ab} = V_{an} - V_{bn}$$

$$V_{bc} = V_{bn} - V_{cn}$$

$$V_{ca} = V_{cn} - V_{an}$$

Inserting the line-to-neutral voltages for a positive phase sequence into the line-to-line equations, we have,

$$V_{ab} = V_{an} - V_{bn} = V_{rms} \angle 0° - V_{rms} \angle -120°$$

$$= V_{rms} \left[e^{j0°} - e^{-j120°} \right] = V_{rms} \left\{ 1 - \left[\cos(120°) - j\sin(120°) \right] \right\}$$

$$= V_{rms} \left[1 + \frac{1}{2} + j\frac{\sqrt{3}}{2} \right] = V_{rms} \left[\frac{3 + j\sqrt{3}}{2} \right] = \sqrt{3} V_{rms} \left[\frac{\sqrt{3} + j1}{2} \right]$$

$$= \sqrt{3} V_{rms} \left[1 \angle 30° \right]$$

$$= \sqrt{3} V_{rms} \angle 30°$$

$$V_{bc} = V_{bn} - V_{cn} = V_{rms} \angle -120° - V_{rms} \angle 120°$$

$$= V_{rms} \left[e^{-j120°} - e^{j120°} \right]$$

$$= V_{rms} \left\{ \left[\cos(120°) - j\sin(120°) \right] \right\} - \left[\cos(120°) + j\sin(120°) \right]$$

$$= V_{rms} \left[-2j\sin(120°) \right] = V_{rms} \left[-2j\frac{\sqrt{3}}{2} \right] = \sqrt{3} \, V_{rms} \left[1 \angle -90° \right]$$

$$= \sqrt{3} \, V_{rms} \angle -90°$$

$$V_{ca} = V_{cn} - V_{an} = V_{rms} \angle 120° - V_{rms} \angle 0°$$

$$= V_{rms} \left[e^{j120°} - e^{j0°} \right] = V_{rms} \left\{ \left[\cos(120°) + j\sin(120°) - 1 \right] \right\}$$

$$= V_{rms} \left[-\frac{1}{2} + j\frac{\sqrt{3}}{2} - 1 \right] = V_{rms} \left[\frac{-3 + j\sqrt{3}}{2} \right] = \sqrt{3} \, V_{rms} \left[\frac{-\sqrt{3} + j1}{2} \right]$$

$$= \sqrt{3} \, V_{rms} \left[1 \angle 150° \right]$$

$$= \sqrt{3} \, V_{rms} \angle 150°$$

If we compare the line-to-neutral voltages with the line-to-line voltages, we have the following relationships.

Line to Neutral Voltages

$$V_{an} = V_{rms} \angle 0°$$

$$V_{bn} = V_{rms} \angle -120°$$

$$V_{cn} = V_{rms} \angle 120°$$

Line to Line Voltages

$$V_{ab} = \sqrt{3} V_{rms} \angle 30°$$

$$V_{bc} = \sqrt{3} V_{rms} \angle -90°$$

$$V_{ca} = \sqrt{3} V_{rms} \angle 150°$$

Line-to-line Voltages in Terms of Line-to-neutral Voltages

$$V_{ab} = \sqrt{3}V_{an}e^{j30°}$$

$$V_{bc} = \sqrt{3}V_{bn}e^{j\,30°}$$

$$V_{ca} = \sqrt{3}\,V_{cn}e^{j30°}$$

The above equations show that the magnitudes of the line-to-line voltages in a balanced three-phase system with a positive phase sequence are 3 times the corresponding line-to-neutral voltages and lead these voltages by 30°.

Instantaneous Three Phase Supply

An armature of an alternator is divided into three groups as shown in the below figure. The coils are named as R_1-R_2, Y_1-Y_2 and B_1-B_2 which is mounted on the same shaft. The ends of each coil are brought out through the slipping and brush arrangement is used to collect the induced e.m.f.

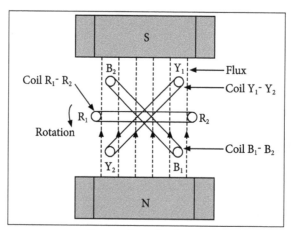

Generation of 3 phase.

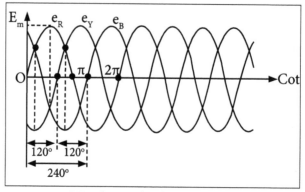

Waveforms of 3 phase voltages.

Let e_R, e_Y and e_B be the three independent voltages induced in coils R_1-R_2, Y_1-Y_2 and B_1-B_2. Alternating voltages have same magnitude and frequency as they are rotated at uniform speed. All of them will be displaced from one other by 120°.

e_R is assumed to be the reference and it is zero for the instant shown in the above figure. At the same instant, e_Y will be displaced by 120° from e_R and will follow e_R while e_B is ahead of e_R by an angle of 120° i.e., if e_R is the reference, then e_Y will attain its maximum and minimum position of 120° later than e_R and e_R will attain its maximum and minimum position at 120° later than e_Y i.e., 120°+120° = 240° later with respect to e_R. All coils together represent the three phase supply system.

The equations for the induced voltages are given as,

$$e_R = E_m \sin(\omega t)$$

$$e_Y = E_m \sin(\omega t - 120°)$$

$$e_B = E_m \sin(\omega t - 240°) = E_m \sin(\omega t + 120°)$$

The waveforms are shown in the above figure.

The phasor diagram of these voltages are shown in the below figure. As phasors rotate in anticlockwise direction, eY lags eR by 120° and eB lags eY by 120°.

If we add three vector ally, then the sum of the three voltages at any instant is zero.

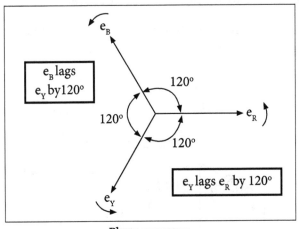

Phase sequence.

$$e_R + e_Y + e_B$$

$$= E_m \sin \omega t + E_m \sin(\omega t - 120°) + E_m \sin(\omega t + 120°)$$

$$= E_m \left[\sin \omega t + \sin \omega t \cos 120° - \cos \omega t \sin 120° + \sin \omega t \cos 120° + \cos \omega t \sin 120° \right]$$

$$=E_m\left[\sin\omega t+2\sin\omega t\,\cos120°\right]=E_m\left[\sin\omega t+2\sin\omega t\left(\frac{-1}{2}\right)\right]=0$$

$$\therefore\qquad\overline{e}_R+\overline{e}_Y+\overline{e}_B=0$$

Phasor addition of all the phase voltages at any instant in the three phase system is always zero.

Important Definitions Related to Three Phase System

Symmetrical System: In poly phase system, magnitudes of different alternating voltages are different. A three phase system in which the three voltages are of same magnitude and frequency and displaced from each other by 120° phase angle is defined as symmetrical system.

Phase Sequence: The sequence in which the voltages in three phases reach their maximum positive values is called as phase-sequence(R-Y-B). The phase sequence determines the direction of rotation of an a.c. motors which has parallel operation of alternators.

Three Phase Supply Connections

In single phase system, two wires are sufficient for transmitting voltage to the load i.e., phase and neutral. But in case of three phase system, two ends of each phase i.e., R_1-R_2, Y_1-Y_2 and B_1-B_2 are available to supply voltage to the toad.

If all six terminals are used independently to supply voltage to load as shown in the below figure, then six wires are required which is expensive:

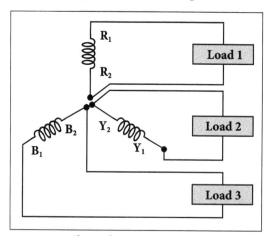

Three phase connections.

To reduce the cost by reducing the number of windings, the three windings are interconnected in a particular fashion which gives three different phase connections.

Star Connection

It is formed by connecting starting or terminating ends of all the three windings together. The ends R_1 -Y_1 - B_1 are connected or ends R_2 -Y_2 - B_2 are connected together. This common point is called as neutral point.

The remaining three ends are brought out for connection purpose. These ends are referred as R-Y-B to which load is connected. The star connection is shown in the below figure:

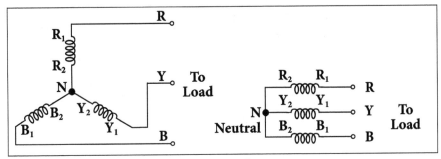

Star connection.

Delta Connection

Delta is formed by connecting one end of winding to starting end of other and connections are continued to form a closed loop. Supply terminals are taken out from the three junction points. The delta connection is shown in the below figure:

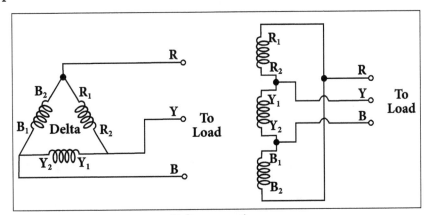

Delta connection.

Relationship between Line and Phase Values of Balanced Star and Delta Connections

Star Connected Load

Let us consider a Y-connected load. We will derive the relationships of voltage, current and power for this connection.

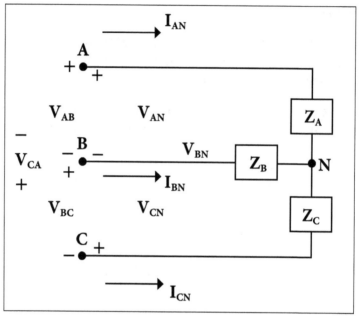

Three phase Y-connection.

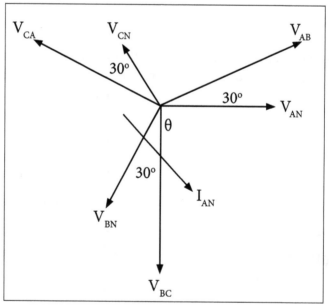

Phasor diagram.

The phase voltages (sequence ABC) are given as,

$$V_{AN} = V_\phi \angle 0°$$

Where,

V_ϕ = Magnitude of phase voltage

$$V_{BN} = V_\phi \angle -120°$$

$$V_{CN} = V_\phi \angle 120°$$

Let us find the line voltages V_{AB}, V_{BC} and V_{CA}.

Using KVL, we have,

$$V_{AB} = V_{AN} + V_{NB}$$

$$= V_{AN} - V_{BN}$$

$$= V_\phi \angle 0° + V_\phi \angle 60° = \sqrt{3}\, V_\phi \angle 30°$$

This can be seen in the above phasor diagram.

Other line voltages are given as,

$$V_{BC} = V_{BN} - V_{CN} = V_\phi \angle -120° - V_\phi \angle 120° = \sqrt{3}V_\phi \angle -90°$$

$$V_{CA} = V_{CN} - V_{AN} = V_\phi \angle 120° - V_\phi \angle 0° = \sqrt{3}V_\phi \angle 150°$$

From the above phasor diagram, we have,

Line voltage = $\sqrt{3}$ Phase voltage

Line current (I_L) = Phase current (I_ϕ)

Line voltage VAB is ahead of phase voltage VAN by 30°.

Total power is given as,

$$P_T = 3 \times \text{Power per phase}$$

$$= 3\left(V_\phi I_\phi \cos\theta\right)$$

$$= \sqrt{3}\, V_L I_L \cos\theta$$

Total reactive power is given as,

$$Q_T = \sqrt{3}\, V_L I_L \sin\theta$$

Apparent power00000 (or VA) = $|S| = \sqrt{P_T^2 + Q_T^2} = \sqrt{3}\, V_L I_L$

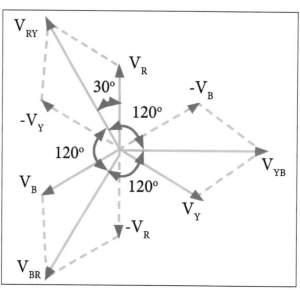

Line voltages and phase voltages in star connection.

It is seen from the above figure that:

- Line voltages are 120° apart from each other.

- Line voltages leads 30° from the corresponding phase voltages.

- The angle φ between line currents and line voltages are (30°+φ) i.e., each line current is lagging (30° + φ) from the corresponding line voltage.

Delta Connected Load

Let us consider a Δ-connected load. The circuit connection and phasor diagram showing the voltages and currents for the balanced circuit is shown in the below figure:

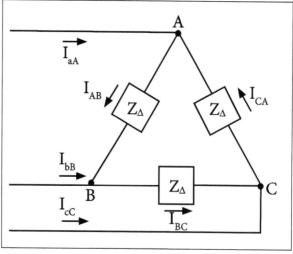

Δ-connected load.

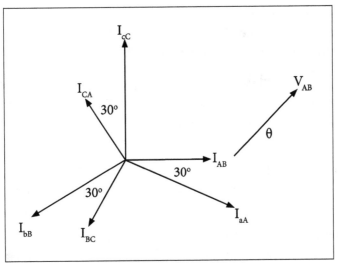

Phasor diagram.

Phase current is given as,

$$I_{AB} = I_{\phi} \angle 0^{\circ}, \ I_{BC} = I_{\phi} \angle -120^{\circ}, \ I_{CA} = I_{\phi} \angle 120^{\circ}$$

$$I_{aA} = I_{AB} - I_{CA} = \sqrt{3} I_{\phi} \angle -30^{\circ}$$

$$I_{bB} = \sqrt{3} I_{\phi} \angle -150^{\circ}, \ I_{cC} = \sqrt{3} I_{\phi} \angle 90^{\circ}$$

For the Δ connected three phase system, we have,

Line current $= \sqrt{3}$ Phase current $\left(I_{\phi} \right)$

Line voltage = Phase voltage

Line current $\left(I_{aA} \right)$ is behind the phase current $\left(I_{AB} \right)$ by 300.

Total power $\left(P_{T} \right) = 3$ Power per phase

$$= 3 \left(V_{\phi} I_{\phi} \cos\theta \right)$$

$$= \sqrt{3} \ V_{L} I_{L} \cos\theta$$

Total reactive power, $Q_{T} = \sqrt{3} \ V_{L} I_{L} \sin\theta$

Apparent power (or VA)$= |S| = \sqrt{P_{T}^{2} + Q_{T}^{2}} = \sqrt{3} \ V_{L} I_{L}$

Advantages of Three Phase System

In three phase system, the alternator armature has three windings which produces three independent alternating voltages. The magnitude and frequency of them is equal but they have a phase difference of 120° between each other. Such a three phase system has following advantages over single phase system:

- For transmission and distribution, three phase system needs less copper or less conducting material than single phase system for given volt amperes and voltage rating and transmission is very economical.

- The output of three phase machine is always greater than single phase machine of same size i.e., approximately 1.5 times. For a given size and voltage, a three phase alternator occupies less space and it is inexpensive than single phase that have same rating.

- In single phase system, the instantaneous power is a function of time and it fluctuates with respect to time. This fluctuating power causes considerable vibrations in single phase motors. Hence performance of single phase motors is poor while instantaneous power in symmetrical three phase system is constant.

- It is possible to produce rotating magnetic field with stationary coils by using three phase system. Thus, it states that three phase motors are self-starting.

- Single phase supply is obtained from three phase supply but three phase cannot be obtained from single phase.

- Power factor of single phase motors is poor than three phase motors of same rating.

- For converting machines like rectifiers, the d.c. output voltage is smoother if number of phases are increased.

- Three phase system gives steady output.

Any further increase in number of phases cause a lot of complications. Hence, three phase system is accepted as standard system throughout the world.

2.3   Three Phase Loads: Housing Wiring, Industrial Wiring and Materials of Wiring

Three phase supply the need for three-phase supply or service occurs when heavy equipments are present such as large motors (beyond 5 HP motors) because such large equipments need high starting and running currents.

Large buildings, plants and offices have greater power requirements than the power used in domestic installations. Therefore, they are installed with three phase wiring or three-phase supply.

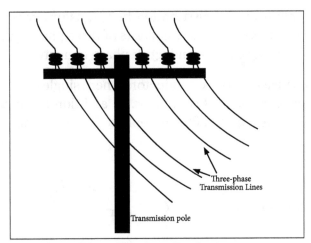

Three phase supply from distribution system.

Three-phase power service is used for high power rated equipments such as large air conditioners, high rated pump sets, air compressors and high torque motors.

Therefore, it is rarely used for domestic installations but commonly used in commercial buildings, offices and industrial installations.

Three Phase AC Supply

Three-phase AC power is generated by a three-phase alternator (also called as AC generators) in the power plants.

In the alternator, three stator windings are separated by some number of degree of rotation and hence the current produced by that coils is also separated by some degrees of rotation which is typically 120 degrees.

This three phase power from the alternators is transmitted to the distribution end through transmission lines.

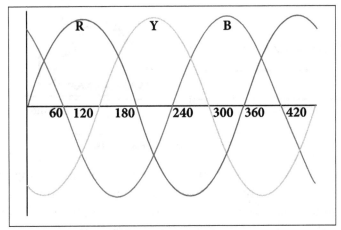

Three phase supply.

The three phase supply from the distribution line transformer is given to the home. Most industrial and commercial services consists of three phase systems that are operated typically at 415V phase to phase and 230V phase to neutral.

Three phase system consists of three conductors where single conductor is used in single phase system excluding neutral conductor. In addition to the three phases, additional neutral conductor is required for three-phase four wire system.

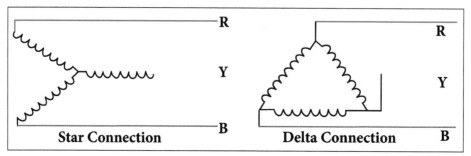

Three-phase three wire systems.

Three-phase systems can be three-phase three wire or three-phase four wire systems. Three-phase 3 line connection consists of three phase conductors and it is employed only where there is no requirement for connecting phase to neutral loads.

These connections can be star or delta depending on the secondary winding of the distribution transformer.

Three-phase 4 wire system is most commonly used connection which consists of three phase conductors and one neutral conductor.

In this three phase wiring, lighting, small-appliance loads and receptacles are connected between phase and neutral while larger equipments such as air conditioners and electric heaters are connected between two phases (i.e., phase to phase).

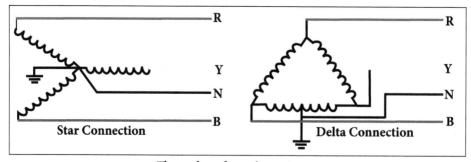

Three phase four wire systems.

Mostly three-phase 4 wire star connection is preferred for connecting both single phase and three phase loads efficiently in a balanced manner. This connection allows phase to neutral connection for small loads. Three phase 4 wire delta connection is used where the phase to neutral load is very small compared with three phase load.

Three phase circuits provide square root of 3 (1.732) times more power compared to single phase power with same current. Thus the three phase system saves electrical installation cost by reducing cable size and size of associated electrical devices.

We can notice the three phase circuits by looking at power line while travelling on roads. Three phase services have a wide range of application in large hotels, restaurants, most factories, office buildings and grocery stores with heavy refrigeration systems.

2.3.1 Housing Wiring

Electricity distribution authorities supply power to the consumers at the following two voltages. Single-phase supply: 230 V, 50 Hz, two-wire and three-phase supply: 415 V, 50 HZ, four-wire. In a two-wire single-phase supply, there is one phase or live wire and the other is called as the neutral wire. Single-phase supply is required for electrical appliances like fan, tube light, lamp, washing machine, refrigerator, electric iron, room heater, room air-conditioner, kitchen electrical appliances like mixer, grinder, microwave oven, etc.,

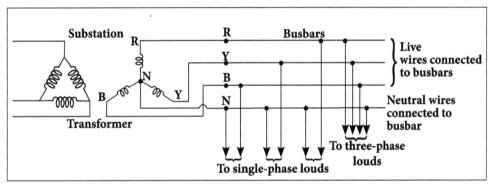

Three phase four-wire distribution system.

In a four-wire three-phase supply, power is supplied through three live wires and a neutral wire. The neutral wire is normally at zero potential and it is earthed at the substation. Three-phase loads like three-phase induction motor is used for water lifting with three-phase supply where all the phases are equally loaded.

Three-phase supply is used to feed single-phase loads as shown in the below figure. Single-phase electrical loads are connected between the phase or live wires and the neutral wire where all the phases are equally loaded.

As shown in the below figure, supply from the secondary of a three-phase transformer is taken through feeder wires to the bus bars. From the bus bars, both single-phase supply and three-phase supply are taken out and are connected to the loads.

2.3.2 Industrial Wiring

Industries or factories are installed with three phase power to connect heavy machineries and equipments. Bus bars carries this three phase power to individual loads through cables. The below figure shows the schematic diagram of industrial three phase wiring:

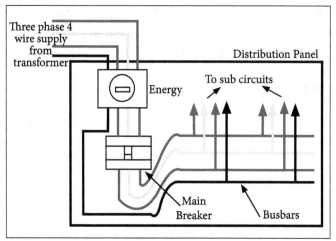

Three phase distribution panel.

Three phase power from the utilities is connected to the main breaker through three-phase energy meter. The power in the main breaker is then given to various bus bars.

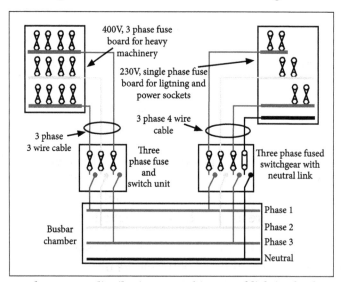

3 phase power distribution to machinery and lighting loads.

This panel has metering arrangement to display parameters like current, voltage, energy and power. The below figure shows the power distribution from main panel to machinery and lighting loads.

Power from the main distribution board is distributed to heavy machinery equipments as well as to lighting boards with power sockets. The power distributed through single and three phase sub-meters is shown in the below figure.

The three phase power distribution to homes or offices is necessary if the load requirement cannot be handled by a single phase supply. The efficient usage of three phase power depends on balancing load distribution on each phase of the three phase supply.

Single phase loads in the offices or homes must be connected to each phase such that maximum possible load balancing is achieved.

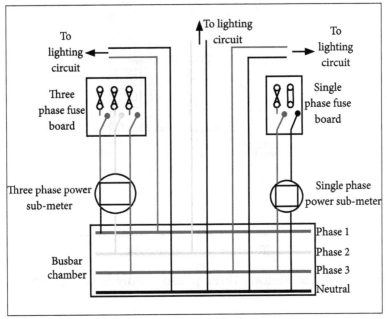

3 phase power distribution to lighting circuits.

The main components in the three-phase wiring to home or building or office premises are shown in the below figure.

The service entrance conductors are connected to a three phase entrance panel. This panel has a three phase main breaker or it has three separate cartridge fuses. This three phase breaker consists of three input lugs to energize three vertical bus bars. This main breaker has single handle such that all the loads are powered down and in electrical faults, it trips or opens all loads simultaneously.

The power from this main panel is connected to the branch circuits. The main panel consists of single pole or double pole or triple pole breakers where phase to ground, phase to phase or three phase loads are connected.

In the above figure, power from the utility pole is connected to sub-circuits via three phase energy meter, three phase breaker (3-pole 60A), double pole RCD, double pole MCB and single pole MCB's.

Single phase and three phase loads connection to three phase power supply is shown in the below figure. The single phase load is connected to the three phase sub-circuits via switches or MCB's.

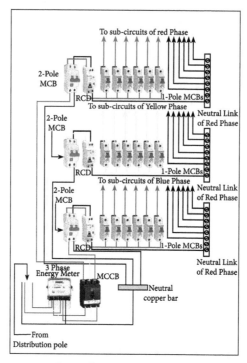

Three phase wiring to home.

For three phase loads like motors, three phase supply is connected via contactor or breaker arrangement.

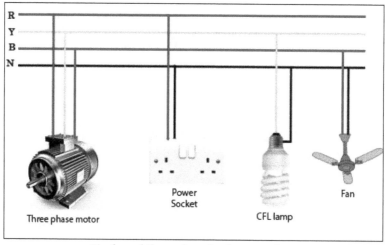

Single and three phase load connection.

A three pole breaker with an appropriate current rating is used for connecting a three phase motor. Proper care should be taken while connecting three phase wires to the motor because the direction of rotation can be reversed by reversing any of the two wires of three phase system.

The wiring diagram for connecting three phase motor to the supply along with control wiring is shown in the below figure. This is a start-stop push button control which includes contactor (M), overload relay, control transformer and push buttons.

The contactor contains large load contacts which can handle large amount of current. The overload relays protect the motor from overload condition by disconnecting power to the coil of the contactor.

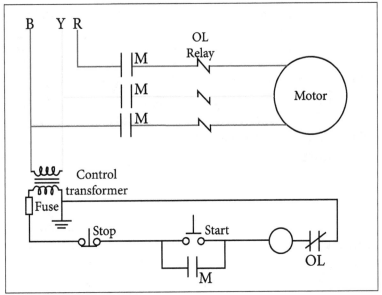

Three phase motor wiring.

The above mentioned information and diagrams gives the basic understanding of three phase power supply distribution to homes and industries.

2.3.3 Materials of Wiring

Modern non-metallic sheathed cables such as (US and Canadian types) NMB and NMC consist of two to four wires which are covered with thermoplastic insulation plus a bare wire for grounding (bonding) which is surrounded by a flexible plastic jacket. Some versions wrap the individual conductors in paper before the plastic jacket is applied.

Special versions of non-metallic sheathed cables such as US type UF are designed for direct underground burial (often with separate mechanical protection) or exterior use where exposure to ultraviolet radiation (UV) is a possible. These cables differ in moisture-resistant construction, lacking paper or other absorbent fillers and it is formulated for UV resistance.

Rubber-like synthetic polymer insulation is used in industrial cables and power cables that are installed underground because of its superior moisture resistance.

Insulated cables are rated by their allowable operating voltage and their maximum operating temperature at the surface of the conductor. A cable may carry multiple usage ratings for applications. For example, one rating for dry installations and another rating when exposed to moisture or oil.

Single conductor building wires are small sized solid wire since the wiring is not very flexible. Building wire conductors larger than 10 AWG (or about 6 mm^2) are stranded for flexibility during installation.

Cables for industrial, commercial and apartment buildings contain many insulated conductors in an overall jacket with helical tape steel or aluminium armour or steel wire armour and lead jacket is used for protection from moisture and physical damage.

Cables intended for very flexible service or in marine applications is protected by woven bronze wires. Power or communications cables (e.g., computer networking) that are routed in or through air-handling spaces (plenums) of office buildings are required under the model building code which is either encased in metal conduit or rated for low flame and smoke production.

Copper Sheathed Mineral Insulated Cables at a Panel Board

For some industrial uses in steel mills and similar hot environments, organic material does not give satisfactory service. Cables insulated with compressed mica flakes are used. Another form of high-temperature cable is a mineral insulated cable where individual conductors are placed within a copper tube and the space is filled with magnesium oxide powder.

The whole assembly is drawn down to smaller sizes by compressing the powder. Such cables have a certified fire resistance rating and they are more costly than non-fire rated cable. They have little flexibility and behave like rigid conduit rather than flexible cables.

Copper Conductors

Electrical devices contain copper conductors because of their multiple beneficial properties including their high electrical conductivity, tensile strength, ductility, creep resistance, corrosion resistance, thermal conductivity, coefficient of thermal expansion, solder ability, resistance to electrical overloads, compatibility with electrical insulators and ease of installation.

Despite competition from other materials, copper is the preferred electrical conductor in all categories of electrical wiring. For example, copper is used to conduct electricity in high, medium and low voltage power networks including power generation, power transmission, power distribution, telecommunications, electronics circuitry, data processing, instrumentation, appliances, entertainment systems, motors, transformers, heavy industrial machinery and other types of electrical equipment.

Aluminium Conductors

Terminal blocks are used for joining aluminium and copper conductors. The terminal blocks are mounted on a DIN rail.

Aluminium wire was common in North American residential wiring from the late 1960's to mid-1970's due to the rising cost of copper. Because of its greater resistivity, aluminium wiring requires larger conductors than copper. Instead of 14 AWG (American wire gauge) copper wire, aluminium wiring need 12 AWG on a typical 15 ampere lighting circuit though local building codes vary.

Solid aluminum conductors were originally made in the 1960's from a utility grade aluminum alloy that had undesirable properties for a building wire and it were used with wiring devices which are intended for copper conductors. These practices causes defective connections and potential fire hazards.

In the early 1970's, new aluminum wire was made from one of the several special alloys which is found in breakers, switches, receptacles, splice connectors, wire nuts etc.,

Unlike copper, aluminium has a tendency to creep or cold-flow under pressure. Newer electrical devices designed for aluminum conductors will compensate this effect.

Unlike copper, aluminium forms an insulating oxide layer on the surface. This is addressed by coating aluminium conductors with an antioxidant paste (containing zinc dust in a low-residue poly butane base [14]) at joints or by applying a mechanical termination which is designed to break through the oxide layer during installation.

Aluminium conductors are used for bulk power distribution and large feeder circuits with heavy current loads. Aluminium cost less than copper and weigh less where larger cross sectional area (i.e., lower resistance and higher mechanical strength) is used for the same weight and price. Aluminium conductors must be installed with compatible connectors and special care must be taken to ensure that the contact surface does not oxidize.

3

Electrical Machines

3.1   Principles of Operation and Characteristics of DC Machines

DC Machines

They are the electro mechanical energy converters which work from a d.c. source and generate mechanical power or convert mechanical power into a d.c. power. These machines are classified into two types on the basis of their magnetic structure.

They are:

- Homo polar machines.

- Hetero polar machines.

Homo Polar Generators: Even though the magnetic poles occur in pairs, in a homo polar generator, the conductors are arranged in such a manner that they always move under one polarity. Either north pole or south pole is used for this purpose. Since the conductor encounters the magnetic flux of the same polarity, it is called as a homo polar generator.

Hetero Polar Generators: In hetero-polar generator, the induced emf in a conductor goes through a cyclic change in voltage as it passes under north and south-pole polarity alternately. The induced emf in the conductor is not a constant but alternates in magnitude.

Working Principle of DC Machine as a Generator

According to Faraday's laws of electromagnetic induction, whenever a conductor is placed in a varying magnetic field or a conductor is moved in a magnetic field, an emf is induced in the conductor. The magnitude of induced emf will be calculated from the emf equation of DC generator. If the conductor is provided with the closed path, then induced current will circulate within the path.

In a DC generator, field coils produces an electromagnetic field and armature conductors are rotated in the field. So, an electromagnetically induced emf is generated in the armature conductors. The direction of induced current is given by Fleming's right hand rule.

According to Fleming's right hand rule, the direction of induced current changes whenever the direction of motion of the conductor changes. Let us consider an armature rotating clockwise and a conductor at the left is moving upward. When the armature completes a half rotation, the direction of motion of that particular conductor will be reversed downward. So, the direction of current in every armature conductor is alternating.

In the below figure, the direction of the induced current is alternating in an armature conductor. In a split ring commutator, connections of the armature conductors is reversed when the current reversal occurs and we get an unidirectional current at the terminals.

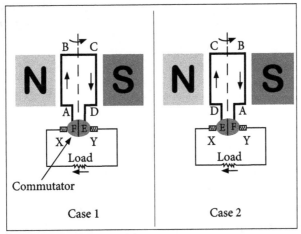

DC machine as generator.

Constructional Features of DC Machines

DC machines are brought out by half cross-sectional views of figures where all the important machine parts are named. Both the small and large industrial machines have the conventional hetero polar cylindrical rotor structure.

The magnetic circuit of a DC machine consists of an armature magnetic material, air gap, field poles and yoke. The yoke of a DC machine is an annular ring. The inter-poles or commutation poles are narrow poles which is fixed to the yoke midway between the main field poles.

An electric field winding, supplies electric energy to establish a magnetic field in the magnetic circuit which results in the great diversity and variety of performance characteristics. Armature winding is connected to the external power source through the commutator brush system which is a mechanical rectifying device for converting the alternating currents.

The below figure shows a single commutator segment and the cross-sectional view of a built-up commutator. The cylindrical rotor or armature of a DC machine is mounted on a shaft which is supported on bearings. One or both the ends of the shaft act as an input or output terminal of the machine and it is coupled mechanically to a load or to a prime mover.

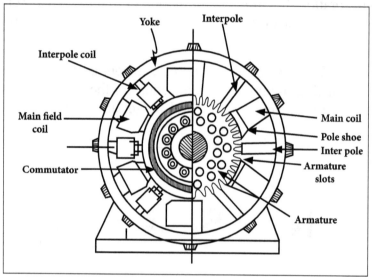

DC machine.

Parallel sided axial slots are used on the rotor surface. In these slots, armature coils are laid as per winding rules. The magnetic material between slots are the teeth. The teeth cross-section influences the performance characteristics of the machine and parameters such as armature coil inductance, magnetic saturation in teeth, eddy-current loss in the stator poles and the cost and complexity of laying armature winding.

DC generator has the following parts:

- Yoke

- Pole cores

- Pole shoes

- Field coils

Yoke: It is the outer cover of DC motor which is also known as frame. It gives protection to the rotating and other part of the machine from moisture, dust etc., Yoke is an iron body which gives the path for the flux to complete the magnetic circuit.

It gives the mechanical support for the poles. Various materials that are used are low reluctance material like cast iron, silicon steel, rolled steel, cast steel etc.,

Purpose of yoke:

- It act as a protecting cover for the whole machine.

- It gives mechanical support for poles.

- It carries the magnetic flux produced by poles.

The field magnets consists of pole cores and pole shoes. The pole shoes serve two purposes as follows:

- They spread out the flux in air gap.

- They support the exciting coils.

Poles and Pole Core: Poles are electromagnetic in nature and the field winding is wound over it. The construction of pole is done using the lamination of particular shape to reduce the power loss due to eddy current. It produces the magnetic flux once the field winding is excited.

Pole Shoe: Pole shoe is an extended part of a pole. It enlarges the area of the pole and more flux will pass through the air gap to armature.

Low reluctance magnetic material like cast steel or cast iron is used for construction of pole and pole shoe.

Armature Core: It is a cylindrical core and it is impregnated. It is made up of high permeability silicon steel stampings. The lamination of the armature core reduces the eddy current loss of the coil.

Commutator: To keep the torque on a DC motor from reversing, the coil moves through the plane which is perpendicular to the magnetic field where a split ring device called a commutator is used to reverse the current at that point. The electrical contacts to the rotating ring are called as brushes since the copper brush contacts were used in early motors.

Functions of commutator:

- It is used to produce unidirectional torque and it is made up of copper and insulating material between the segments is mica.

- It converts an AC emf generated internally into DC emf.

Carbon Brush

The brushes of DC motor are made up of carbon or graphite structures which makes sliding contact over the rotating commutator. The brushes are used to relay the current from external circuit to the rotating commutator from where it flows into the armature winding. Thus, the commutator and brush unit of the DC motor transmits the power from the static electrical circuit to the mechanically rotating region or the rotor.

3.2   Transformers

Single Phase Transformer

A transformer is defined as a static device which is used to transform electric power in one circuit to electric power of the same frequency in another circuit. The voltage can be raised or lowered in a circuit with a proportional increase or decrease in the current ratings.

The main principle of operation of a transformer is the mutual inductance between two circuits which is linked by a common magnetic flux.

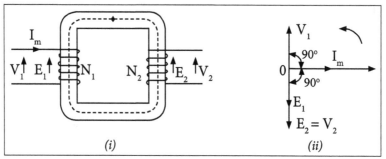

(i) Single phase transformer (ii) Pharos diagram.

A basic transformer consists of two coils which are electrically separated and inductive but are magnetically linked through the path of reluctance.

An ideal transformer has the following properties:

- It has no leakage flux i.e., the same flux links both the primary and the secondary winding.

- It has no winding resistance.

- There is no iron losses in the core.

Copper loss in the transformer depends on current and iron loss depends on the voltage. Thus, the total loss in the transformer depends on the volt-ampere product and not on the phase angle between the voltage and the current. i.e., it is independent of load power factor and the rating of a transformer is in kVA but not in kW.

Construction of Transformer

All transformers have the following essential parts:

- Two or more electrical windings are insulated from each other and from the core.

- A core in a single phase distribution transformers comprises of cold rolled silicon steel strip instead of an assembly of punched silicon steel laminations such as those used in the larger power transformer cores.

The flux path in the assembled core is parallel to the direction of the orientation of the steel. This reduces the core losses for a given flux density.

Other necessary parts are as follows:

- A suitable container for the assembled core and windings.

- A suitable medium for insulating the core and its windings from each other and from the container.

- Suitable bushings for insulating and bringing the terminals of the windings out of the case.

The two basic types of transformer construction are as follows:

- Core type.

- Shell type.

In the core type transformer, the copper surrounds the iron core while in the shell type, the iron surrounds the copper winding.

Core Type Transformer

The magnetic circuit of this transformer is in the shape of the hollow rectangle as shown in the below figure in which I_0 is the no-load current and φ is the flux produced by it. N_1 and N_2 are the number of turns on the primary and secondary sides.

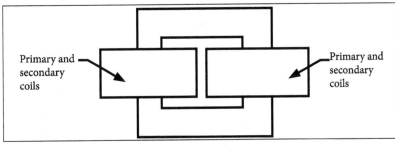

Core type transformer.

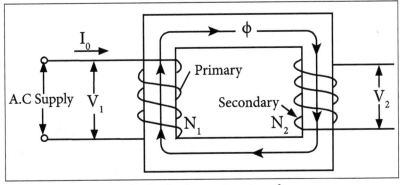

Magnetic circuit of a core type transformer.

The two coils namely high and low side coils are made up of silicon steel laminations which are either rectangular or L-shaped. To provide maximum linkage between windings, the group or each leg is made up of both high tension and low tension coils.

If the high voltage coils were adjacent to the core, an additional high voltage insulation layer is required between the coils and the iron core. Rectangular cores with rectangular cylindrical coils is used for small size core-type transformers. Circular cylindrical coils are preferred with square cores for large size transformers.

A common improvement on the square core uses a cruciform core which demands at least two sizes of core strips. Core stepping is done for very large transformers where three stepped core are used. Core stepping reduces the length of the mean turn and I2R loss.

Shell Type Transformers

In the shell-type construction, the iron surrounds the copper. The core is made up of E-shaped or F-shaped laminations which are stacked to give a rectangular figure. All the windings are placed on the center leg and to reduce leakage, each high side coil is adjacent to a low side coil.

The coils occupy the entire space of both windows that are flat in shape and they are constructed with strip copper. To reduce the amount, high voltage insulation is required where the low voltage coils are placed adjacent to the iron core.

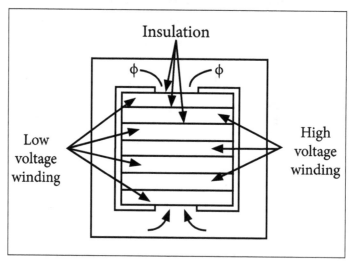

Shell type transformer.

Core type is better adopted for some high voltage service since there is more space for insulation. The shell type has better provision for mechanically supporting and bracing the coils. This allows better resistance to high mechanical forces that develop during a high current short circuit. Both the core and shell forms are used and selection is based on many factors such as voltage rating, KVA rating, weight, insulation stress, mechanical stress and heat distribution.

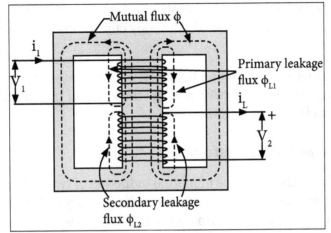

Magnetic path of shell type transformer.

Construction of Transformer (Single Phase)

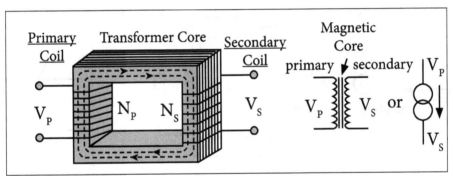

Construction of transformer and its symbol.

V_P = Primary voltage

V_S = Secondary voltage

N_P = Number of primary windings

N_S = Number of secondary windings

ϕ = Flux linkage

Transformer has two coils which have mutual inductance and a laminated steel core. The two coils are insulated from each other and from the steel core. It needs some suitable container for the assembled core and windings which are insulated.

In order to insulate and to bring out the terminals of the winding from the tank, bushings from either porcelain or capacitor type must be used.

In all transformers, the core is made up of transformer sheet steel laminations that are assembled to provide a continuous magnetic path with a minimum air gap. The steel should have high permeability and low hysteresis loss. The steel is made up of high silicon content and it must be heat treated. The eddy current loss is reduced by laminating the core.

Lamination is done with the help of a light coat of core plate varnish or an oxide layer on the surface for a frequency of 50Hz and the thickness of the lamination varies from 0.35 mm to 0.5 mm for a frequency of 25Hz.

Working Principle of Transformer

Transformer works on the principle of mutual induction of two coils or Faraday law's of electromagnetic induction. When the current in the primary coil is changed, flux linked to the secondary coil also changes. An emf is induced in the secondary coil due to the Faraday law's of electromagnetic induction.

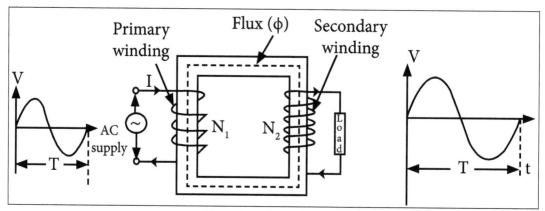

Operation of transformer.

Transformer consists of two inductive coils which are electrically separated but linked through a common magnetic circuit. The two coils have high mutual inductance. One of the two coils is connected to a source of alternating voltage and it is said to be primary winding and another winding is connected to load which os said to be secondary winding.

The primary winding has N_1 number of turns while the secondary winding has N_2 number of turns. When primary winding is excited by an alternating voltage, it circulates an alternating current which in turn produces an alternating flux that completes its path through common magnetic core. Thus, an alternating flux links with the secondary winding.

As the flux is alternating, according to the Faraday's law of electromagnetic induction, mutually induced emf is developed in the secondary winding. If the load is connected to the secondary winding, emf drives a current through it.

Transformers can be categorized in different ways depending on their purpose, use, construction etc., such as follows.

Step up Transformer and Step Down Transformer: They are used for stepping up and stepping down the voltage level of power in the transmission and distribution power system network.

Three Phase Transformer and Single Phase Transformer: Former is used in three phase power system as it is cost effective than latter. But when the size matters, it is preferable to use a bank of three single phase transformer as it is easier to transport than one single three phase transformer unit.

Electrical Power Transformer, Distribution Transformer & Instrument Transformer: Power transformers are used in transmission network for stepping up or down the voltage level. It operates during high or peak loads and it has maximum efficiency at or near full load.

Distribution transformer steps down the voltage for the distribution purpose to domestic or commercial users. It has good voltage regulation and operates 24 hours a day with maximum efficiency at 50% of full load.

Instrument transformers include C.T and P.T which are used to reduce high voltages and the current to lesser values.

Two Winding Transformer and Auto Transformer: Former is used where ratio between high voltage and low voltage is greater than 2. It is cost effective to use latter where the ratio between high voltage and low voltage is less than 2.

Oil Cooled and Dry Type Transformer: In oil cooled transformer, the cooling medium is transformer oil, whereas, dry type transformer is air cooled.

Outdoor Transformer and Indoor Transformer: Transformers which are designed for installing at outdoor are outdoor transformers and transformers which are designed for installing at indoor are indoor transformers.

Emf Equation of a Transformer and Voltage Transformation Ratio

In a transformer, the source of alternating current is applied to the primary winding. Due to this, the current in the primary winding produces alternating flux in the core of a transformer. This alternating flux is linked with the secondary winding and due to mutual induction; an emf is induced in the secondary winding. Magnitude of this induced emf is found by using the following emf equation of the transformer.

N_1 = Number of turns in primary winding

N_2 = Number of turns in secondary winding

ϕ_m = Maximum flux in the core $(\text{Wb}) = (B_m \times A)$

f = Frequency of an AC supply (Hz)

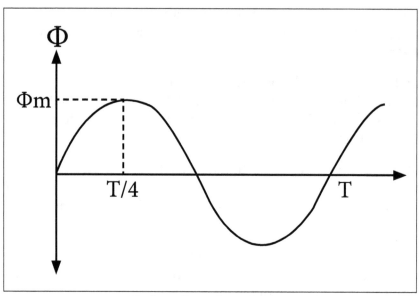

Emf waveform of transformer.

As shown in the above figure, the flux rises sinusoid ally to its maximum value φ_m from 0. It reaches to the maximum value in one quarter of the cycle i.e., in T/4 sec (T is the time period of the sine wave of the supply = 1/f).

Average rate of change of flux $= \phi_m / (T/4) = \phi_m / (1/4f)$

Average rate of change of flux $= 4f\ \phi_m .... (Wb/s)$

Induced emf per turn = Rate of change of flux per turn,

Average emf per turn $= 4f\ \phi_m ...(Volts)$

Form factor = RMS value/Average value

RMS value of emf per turn = Form factor X Average emf per turn

Since the flux φ varies sinusoid ally, form factor of a sine wave is 1.11.

RMS value of emf per turn $= 1.11 \times 4f\phi_m = 4.44\,f\phi_m$

RMS value of induced emf in whole primary winding (E_1) = RMS value of emf per turn x Number of turns in primary winding

$$E_1 = 4.44f\ N_1\ \phi_m \qquad ...(1)$$

RMS value of an induced emf in the secondary winding (E_2) is given as,

$$E_2 = 4.44f\ N_2\ \phi_m \qquad ...(2)$$

From the above equations (1) and (2), we have,

$$\frac{E_1}{N_1} = \frac{E_2}{N_2} = 4.44 f \, \Phi m$$

This is called as the emf equation of transformer where emf/number of turns is same for both primary and secondary winding.

For an ideal transformer on no load, we have,

$$E_1 = V_1$$

$$E_2 = V_2$$

Where,

V_1 = Supply voltage of primary winding.

V_2 = Terminal voltage of secondary winding.

Voltage Transformation Ratio (K)

$$\frac{E_1}{N_1} = \frac{E_2}{N_2} = K$$

Where,

K = Constant = Voltage transformation ratio

If $N_2 > N_1$, i.e., $K > 1$ then the transformer is called as step-up transformer.

If $N_2 < N_1$, i.e., $K < 1$ then the transformer is called as step-down transformer.

Equivalent Circuit of Transformer

Equivalent circuit diagram of any device is used in the pre-determination of the behavior of the device under the various condition of operation.

The simplified equivalent circuit of a transformer is drawn by representing all parameters of the transformer either on the secondary side or on the primary side. The equivalent circuit diagram of transformer is shown in the below figure.

The transformation ratio of an equivalent circuit of a transformer is $K = E_2/E_1$. The induced emf E_1 is equal to the primary applied voltage V_1 minus primary voltage drop. This voltage introduces no load current (I_o) in the primary winding of the transformer. The value of no load current is very small and it can be neglected. Hence, $I_1 = I_1'$. The no load current is divided into two components called as the magnetizing current (I_m) and working current (I_w).

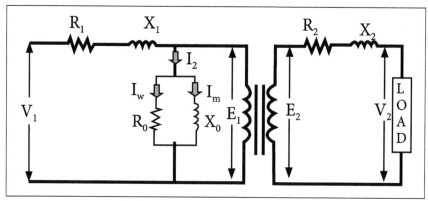

Equivalent circuit diagram of a transformer.

These two components of no load current are due to current drawn by a non-inductive resistance R_0 and pure reactance X_0 with voltage E_1.

Secondary current I_2 is given as,

$$I_2 = \frac{I_1'}{K} = \frac{I_1 - I_0}{K}$$

Terminal voltage V_2 across the load is equal to the induced emf E_2 in the secondary winding minus voltage drop in the secondary winding.

Equivalent circuit when all the quantities are referred to primary side is shown in the below figure:

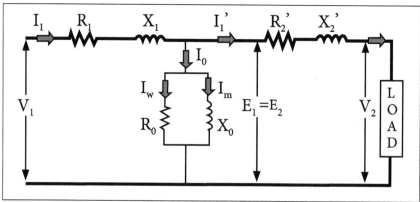

Circuit diagram of transformer when all secondary quantities are referred to the primary side.

Secondary resistance referred to primary side is expressed as,

$$R_2' = \frac{R_2}{K^2}$$

Equivalent resistance referred to primary side is given as,

$$R_{ep} = R_1 + R_2'$$

Secondary reactance referred to primary side is given as,

$$X_2' = \frac{X_2}{K^2}$$

Equivalent reactance referred to primary side is represented as,

$$X_{ep} = X_1 + X_2'$$

Equivalent circuit when all the quantities are referred to secondary side is shown in the below figure.

The primary resistance referred to secondary side is expressed as,

$$R_1' = K_2 R_1$$

The equivalent resistance referred to secondary side is expressed as,

$$R_{es} = R_2 + R_1'$$

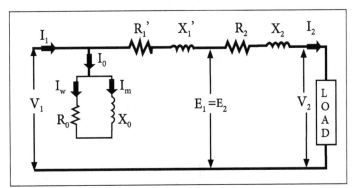

Circuit diagram of transformer when all the primary quantities are referred to secondary side.

Primary reactance referred to secondary side is obtained as,

$$X_1' = K_2 X_1$$

The equivalent reactance referred to secondary side is represented as,

$$X_{eq} = X_2 + X_1'$$

No load current I_o is 3% to 5% of full load rated current and the parallel branch consist of resistance R_o and reactance X_o which is omitted without introducing any appreciable error in the behavior of the transformer under the loaded condition.

Equivalent circuit of the transformer can be done by neglecting the parallel branch consisting of R_o and X_o. The simplified circuit diagram of the transformer is shown in the below figure:

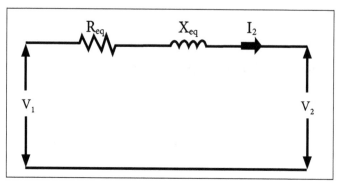

Simplified equivalent circuit diagram of a transformer.

Voltage regulation is the percentage of voltage difference between no load and full load voltages of a transformer with respect to its full load voltage.

Three Phase Transformer

It is used for electrical power generation, transmission and distribution for all industrial uses. Three-phase supplies have many electrical advantages over single-phase power and when considering three-phase transformers, we have to deal with three alternating voltages and currents which differ in phase-time by 120 degrees as shown in the below figure:

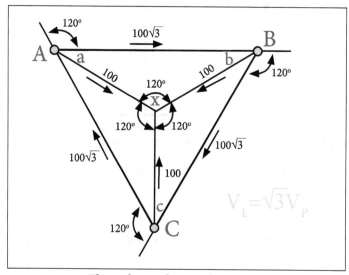

Three phase voltage and currents.

V_L = Line-to-line voltage

V_p = Phase-to-neutral voltage

A transformer cannot act as a phase changing device and it changes a single-phase into three-phase or three-phase into single phase. To make the transformer connections compatible with three-phase supplies, we should connect them together to form a three phase transformer configuration.

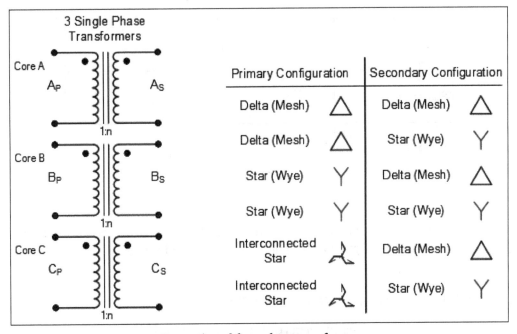

3 Single Phase Transformers	Primary Configuration	Secondary Configuration
	Delta (Mesh) △	Delta (Mesh) △
	Delta (Mesh) △	Star (Wye) Y
	Star (Wye) Y	Delta (Mesh) △
	Star (Wye) Y	Star (Wye) Y
	Interconnected Star	Delta (Mesh) △
	Interconnected Star	Star (Wye) Y

Connection of three phase transformer.

A three phase transformer is constructed either by connecting three single-phase transformers or by using one pre-assembled and balanced three phase transformer which consists of three pairs of single phase windings which is mounted on one single laminated core.

The advantages of building a single three phase transformer is that, for the same kVA rating, it will be smaller, cheaper and lighter than three individual single phase transformers connected together since the copper and iron core are used more effectively. The methods of connecting the primary and secondary windings are the same.

Primary and secondary windings of a transformer is connected in different configuration to meet any requirement. In three phase transformer windings, three forms of connection are possible namely star (wye), delta (mesh) and interconnected-star (zigzag).

Combinations of the three windings with the primary delta-connected and the secondary star-connected or star-delta, star-star or delta-delta depends on the transformers. When transformers provide three or more phases, they are referred to as a poly phase transformer.

Line voltage and current of three phase transformer is shown in the below table:

Primary – Secondary Configuration	Line Voltage Primary or Secondary	Line Current Primary or Secondary
Delta-Delta	$V_L \Rightarrow nV_L$	$I_L \Rightarrow \dfrac{I_L}{n}$
Delta –Star	$V_L \Rightarrow \sqrt{3}.nV_L$	$I_L \Rightarrow \dfrac{I_L}{\sqrt{3}.n}$
Star-Delta	$V_L \Rightarrow \dfrac{nV_L}{\sqrt{3}}$	$I_L \Rightarrow \sqrt{3}.\dfrac{I_L}{n}$
Star-Star	$V_L \Rightarrow n V_L$	$I_L \Rightarrow \dfrac{I_L}{n}$

Problems

1. A 1-phase transformer has 180 turns in its secondary and primary windings. Resistance are 0.233Ω and 0.067Ω. Let us calculate the equivalent resistance of the following:

(i) Primary winding in terms of the secondary winding.

(ii) Secondary winding in terms of the primary winding.

(iii) Total resistance of the transformer in terms of the primary winding.

Solution:

Given:

$$N_1 = N_2 = 180 \text{ turns}$$

$$R_1 = 0.067\Omega$$

$$R_2 = 0233\ \Omega$$

Formula:

$$K = \frac{N_2}{N_1}$$

$$R'_2 = \frac{R_2}{K^2}$$

$$R_{01} = R_1 + R_2'$$

(i) Primary in Terms of the Secondary [Equivalent Resistance]:

$$R_1 = R_1 K^2$$

Transformation ratio is given as,

$$K = \frac{N_2}{N_1}$$

$$= \frac{180}{180} = 1$$

$$\therefore \qquad R_1' = R_1 K^2$$

$$= [0.067 \times 12] = 0.067\,\Omega$$

(ii) Equivalent Resistance of Secondary in Terms of Primary Winding:

$$R_2' = \frac{R_2}{K^2}$$

$$= \frac{0.233}{(1)^2} \Rightarrow 0.233\,\Omega$$

(iii) Total Resistance Referred to Primary Winding:

$$R_{01} = R_1 + R_2'$$

$$= 0.067 + 0.233$$

$$R_{01} = 0.3\,\Omega$$

2. The parameters of approximate equivalent circuit of a 4 KVA, 200/400 V, 50 Hz, 1 transformer are given as follows.

$$R_p' = 0.15\,\Omega$$
$$X_p' = 0.37\,\Omega = 0.3\,\Omega$$
$$R_o = 600\,\Omega$$
$$X_m = 300\,\Omega$$

When a rated voltage of 200 V is applied to the primary, a current of 10 A at lagging power factor of 0.8 flows in the secondary winding. Let us calculate the following:

(1) Current in the primary, I_p.

(2) Terminal voltage at the secondary side.

Solution:

Given:

Capacity $= 4$ KVA

$$R_p = R_{01} = 0.15\Omega$$

$$X_p = X_{01} = 0.37\Omega$$

$$R_0 = 600\Omega$$

$$X_m = X_0 = 300\Omega$$

Formula:

$$I_P = \sqrt{I_0^2 + I_2'^2 + 2I_0\, I_2'\cos\phi}$$

$$I_0 = \sqrt{I_\mu^2 + I_W^2}$$

$$K = \frac{V_2}{V_1}$$

Voltage drop $= I_2\, R_{02}\cos\phi + I_2\, X_{02}\sin\phi$

$$\frac{V_1}{V_2} = 200/400\,V$$

$$I_2 = 10A$$

$$\cos\phi = 0.8\left(\text{Lagging}\right)$$

(1) $I_P = \sqrt{I_0^2 + I_2'^2 + 2I_0\, I_2'\cos\phi}$

$$I_0 = \sqrt{I_\mu^2 + I_W^2}$$

$$I_\mu = \frac{V_1}{X_o} = \frac{200}{300}$$

$$= 0.666A$$

$$I_W = \frac{V_1}{R_o} = \frac{200}{600}$$

$$= 0.333A$$

$$\therefore \quad I_0 = \sqrt{(0.666)^2 + (0.333)^2} = 0.745A.$$

$$\tan\theta = \frac{I_W}{I_\mu} = \frac{0.333}{0.666}$$

$$\tan\theta = 0.5$$

$$\theta = \tan^{-1}(0.5)$$

$$= 26.56°$$

$$= 90° - 26.56 = 63.43° \text{ is the angle between } I_0 \text{ and } V_1$$

$$V_2 = KV_1$$

$$K = \frac{V_2}{V_1}$$

$$= \frac{400}{200} = 2.$$

$$I_2' = 2(10) = 20 \text{ A}$$

$$\therefore \quad I_P = \sqrt{0.745^2 + 20^2 + 2(0.745 \times 20)\cos 26.5°}$$

$$= 20.67 \text{ A}$$

(2) Terminal voltage at secondary side = V_2 - Approximate voltage drop

Approximate voltage drop = $I_2 R_{02}\cos\phi + I_2 X_{02}\sin\phi$

$$R_{02} = K^2 R_{01}$$

$$= (2)^2 \, 0.15 = 0.6\Omega$$

$$X_{02} = K^2 X_{01}$$

$$= (2)^2 \, 0.37 = 1.48\Omega$$

$$\cos \quad 0.8$$

$$= \cos^{-1}(0.8) = 36.86°$$

$$\sin = \sin \, (36.86) = 0.6°$$

Voltage drop $= 10(0.6 \times 0.8) + 10(1.48 \times 0.6)$

$$= 13.7 \text{ V}$$

Terminal voltage at secondary side $= 400 - 13.7$

$$= 386.3 \text{ V}$$

Efficiency of Transformer

In a practical transformer, two types of major losses namely core and copper losses occur. The losses are wasted as heat and temperature.

Output power of the transformer is always less than the input power drawn by the primary from the source and efficiency is defined as,

$$\eta = \frac{\text{Output power in KW}}{\text{Output power in Kw} + \text{Losses}}$$

$$= \frac{\text{Output power in KW}}{\text{Output power in Kw} + \text{Core loss} + \text{Copper loss}}$$

Efficiency of the transformer for general loading is expressed as,

$$\eta = \frac{xS \cos\theta}{xS \cos\theta + P_{core} + x^2 P_{cu}}$$

All Day Efficiency

All day efficiency is also called as energy efficiency. To estimate the efficiency, the whole day is broken into several time blocks over which the load remains constant. The idea is to calculate the total amount of energy output in KWH and the total amount of energy input in KWH over a complete day and then take ratio of these to get the energy efficiency or all day efficiency of the transformer.

Energy or all day efficiency of a transformer is expressed as,

$$\eta_{all\,day} = \frac{\text{Energy output in KWH in 24 hours}}{\text{Energy input in KWH in 24 hours}}$$

$$= \frac{\text{Energy output in KWH in 24 hours}}{\text{Output in KWH in 24 hours + Energy loss in 24 hours}}$$

$$= \frac{\text{Output in KWH in 24 hours}}{\text{Output in KWH in 24 hours + Loss in core in 24 hours}}$$
$$+ \text{Loss in the Winding in 24 hours}$$

$$= \frac{\text{Energy output in KWH in 24 hours}}{\text{Energy output in KWH in 24 hours} + 24P_{core}}$$
$$+ \text{Energy loss(cu)in the winding in 24 hours}$$

3.3   Synchronous Machines

Synchronous machines are AC machines where a field circuit is supplied by an external DC source. It consists of two major parts namely stationary part (stator) and a rotating field (rotor).

Principle of an Alternator

It is known that the electric supply used now-a-days for commercial as well as domestic purposes is of alternating type. The AC machines associated with the alternating voltages are classified as generators and motors.

Machines which are generating AC are called as alternators or synchronous generators, whereas, the machines which are accepting the input from an AC supply to produce mechanical output are called as synchronous motors. Both these machines work at a specific constant speed called as synchronous speed and it is called as synchronous machines.

The working principle of an alternator is very simple which is similar to the basic principle of DC generator. It depends on the Faraday's law of electromagnetic induction which says that the current is induced in the conductor inside a magnetic field when there is a relative motion between that conductor and the magnetic field.

Construction

An alternator mainly consists of two parts as follows:

• Stator

• Rotor

Stator

The armature core is supported by stator frame and it is built up of laminations of special magnetic iron or steel iron alloy where the core is laminated to minimize loss due to eddy currents. The laminations are stamped out in complete rings or segments. The laminations are insulated from one another and the space between them allows the cooling air to pass through.

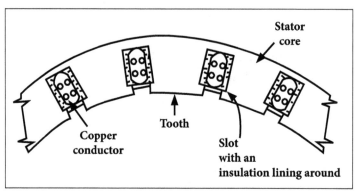

Construction of stator.

The inner periphery of the stator is slotted and the copper conductors are joined to one another where armature winding is housed in the slots. The other ends of the winding are brought out and they are connected to the fixed terminal from which generator power is taken out. Different shapes of the armature slots are shown in the above figure.

The wide open type slot used in DC machines permits easy installation of form-wound coils and easy removal in case of repair but it has disadvantage of distributing the air gaps flux into bunches that produce ripples in wave of generated EMF.

The semi-closed type slots are better which do not use form wound coils. The fully closed slots do not disturb the air gap flux but they increase the inductance of the windings.

The armature conductors are threaded to increase the initial labor and cost of the winding. Hence, they are rarely used.

Rotor

Depending on the type of application, they are classified into two types as follows:

• Salient-pole or the projecting pole type.

• Non salient-pole or round rotor or cylindrical rotor.

Salient pole rotor: This is also called as projected pole type since all the poles are projected out from the surface of the rotor. The poles are built up of thick steel laminations. It is bolted to the rotor as shown in the below figure. Its face has a specific shape. The field winding is provided on the pole shoe. These rotors have large diameters and small axial lengths.

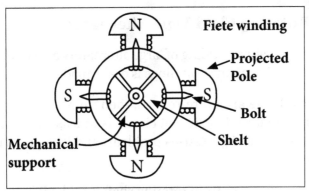

Salient pole type rotor.

The limiting factor for the size of the rotor is the centrifugal force that acts on the rotating member of the machine.

As mechanical strength of salient pole type is less, it is preferred for low speed alternators ranging from 125 r.p.m to 50 r.p.m. The prime movers used to drive such rotors are water turbines and I.C engines.

Smooth cylindrical rotor: This is also called as non-salient type or non-projected pole type or round rotor. It consists of smooth solid steel cylinder which has a number of slots to support the field coil. These slots are covered at the top with the help of steel or manganese wedges.

The un-slotted portions of the cylinder act as the poles. The poles are not projecting out and the surface of the rotor is smooth which help us to maintain an uniform air gap between stator and rotor.

These rotors have small diameters and large axial lengths. This keeps the peripheral speed within the limit. The main advantage of this type is that these are mechanically very strong and preferred for high speed alternators ranging between 1500 to 3000 rpm.

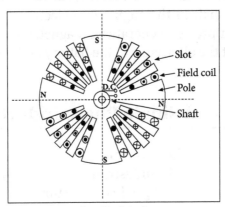

Smooth cylindrical rotor.

High speed alternators are known as turbo-alternators. The prime movers used to drive such type of rotors are steam turbines and electric motors.

Advantages of Rotating Field Type Alternator

Ease of construction: Armature winding of large alternators are complex, so that the connection and bracing of the armature windings is easily made for the stationary stator.

Number of slip rings: If the armature rotates, then number of slip rings required for power transfer from armature to external circuit is at least three. Heavy current that flows through brush and slip rings causes problems and it requires more maintenance in large alternators.

Insulation required for slip rings from rotating shaft is difficult in the rotating armature system.

Better insulation to armature: Insulation arrangement of armature windings is made from core on stator.

Reduced rotor weight and rotor inertia: Insulation requirement is less since the field system is placed on the rotor. Also rotational inertia is less. It takes lesser time to gain full speed.

Improved ventilation arrangement: Cooling occurs when we enlarge the stator core with radial ducts. Water cooling is easier if the armature is housed in the stator. In all the alternators, the armature is housed in the stator while the DC field system is placed in the rotor.

Applications of alternator:

- Alternators are used in modern automobiles to charge the battery and to power the electrical system when its engine is running.

- Alternators provide electricity to power the accessory of the vehicle such as lights and radio and charge its battery. It converts the mechanical energy created by the crankshaft in the engine into electrical energy via induction. The energy-conversion process generates a magnetic field which then charges the wires that connect the alternator to the battery and accessories.

3.4   Three Phase and Single Phase Induction Motors

Three phase induction motor is the most commonly used electrical motor. Almost 80% of the mechanical power used by the industries is provided by using three phase induction motors due to its simple and rugged construction, absence of commentator and good speed regulation, low cost and good operating characteristics.

In three phase induction motor, the power is transferred from the stator to the rotor winding through the induction. The induction motor is also termed as a synchronous motor as it runs at a speed other than the synchronous speed. Three phase induction motor has two main parts as follows:

- Stator
- Rotor

Stator

It consists of a steel frame that supports a hollow and cylindrical core of stacked laminations.

Slots on the internal circumference of the stator poles carries the stator winding. It is a stationary part of an induction motor.

A stator winding is placed in the stator of induction motor and the three phase supply is given to it. The stator of the three phase induction motor consists of three main parts as follows:

- Stator frame.
- Stator core.
- Stator winding or field winding.

Depending on the type of rotor construction, the three phase inductor motor are classified as follows:

- Squirrel cage induction motor.
- Slip ring induction motor or wound induction motor or phase wound induction motor.

The construction of stator for both the types of three phase induction motor remains the same:

- One of the problems with an electrical motor is the production of heat during its rotation. In order to overcome this problem, we need fan for cooling.
- Shaft required for transmitting the torque to the load is made up of steel.
- There is a small distance between rotor and the stator which normally varies from 0.4 mm to 4mm. Such a distance is termed as air gap.
- Bearings are used to support the rotating shaft.
- For receiving external electrical connection, terminal box is needed.

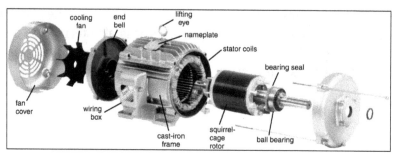

Cross sectional view of 3-phase induction motors.

Stator Frame

It is the outermost part of the three phase induction motor. Its main function is to support the stator core and the field winding. It acts as a covering and it provides protection and mechanical strength to all the inner parts of an induction motor. The frame is either made up of die cast or fabricated steel.

The frame of three phase induction motor is very strong and rigid as the air gap length of three phase induction motor is small. Rotor does not remain concentric with stator which will give rise to an unbalanced magnetic pull.

Stator Core

The main function of the stator core is to carry the alternating flux. In order to reduce the eddy current loss, the stator core is laminated. These laminated types of structure are made up of stamping which is about 0.4 to 0.5 mm thick.

All the stampings are stamped together to form a stator core which is then housed in the stator frame. Stamping is made up of silicon steel which reduces the hysteresis loss in motor.

Stator Winding or Field Winding

The slots on the periphery of stator core of the three phase induction motor carries three phase windings. This three phase winding is supplied by the three phase AC supply. The three phases of the winding are connected either in star or delta depending on the type of starting method.

The squirrel cage motor is started by star-delta stater and the stator of squirrel cage motor is delta connected. The slip ring three phase induction motor are started by inserting resistances since the stator winding of slip ring induction motor is connected either in star or delta.

The winding wound on the stator of three phase induction motor is also called as field winding and when this winding is excited by three phase AC supply, it produces a rotating magnetic field.

Slip-torque Characteristics

Slip speed of an induction motor is the relative speed of rotor with respect to the rotating magnetic field.

It is given as,

$$\text{Slip speed} = N_1 - N_2$$

Where,

$$N_2 = \text{Motor speed.}$$

Slip of a three-phase induction motor is given as,

$$s = \frac{N_1 - N_2}{N_1} \times 100\%$$

Where,

$$N_2 = \text{Speed of the rotor in rpm.}$$

N_1 is the synchronous speed of the rotating magnetic field due to the stator winding which is given as,

$$N_1 = \frac{120 f_1}{P_1}$$

Where,

f_1 = Frequency of the three-phase supply at the stator winding (50 Hz or 60 Hz).

P_1 = Number of magnetic poles in the rotating magnetic field caused by the stator winding.

When the motor is loaded, the rotor tends to retard more and the slip increases. In practical induction motors, typical value of slip at full load is within 5%. When the motor is not loaded, slip is small and it is zero. In that condition $N_2 \approx N_1$.

When the rotor is standstill, $N_2 = 0$, $s = 1$. At the instant of starting, slip is unity. For an induction motor, $0 < s \leq 1$.

Torque -slip Characteristics Curve

The total torque developed is expressed as,

$$T_m = \left(\frac{3}{\omega_s}\right) \cdot \frac{s E_2^2}{\left[R_2^2 + (s X_2)^2\right]} \cdot R_2$$

Let us examine the torque versus speed characteristics for different operating conditions as follows.

Case 1: Motor running near synchronous speed (s is very small)

Case 2: Starting state of motor (s = 1)

Case 1

Torque expression is given as,

$$T_m = \left(\frac{3}{\omega_s}\right) \cdot \frac{s E_2^2}{\left[R_2^2 + (s X_2)^2\right]} \cdot R_2$$

Slip 's' is very small and hence $R_2^2 \gg (s X_2)^2$

Torque expression is given as,

$$T_m = \left(\frac{3}{\omega_s}\right) \cdot \frac{s E_2^2}{R_2}$$

$$T_m \, \alpha \, \frac{s E_2^2}{R_2}$$

Where,

α = Proportionality

Torque increases linearly with slip near synchronous speed. If the rotor resistance is high, then rated torque is reduced. Torque is proportional to the square of the applied voltage.

Case 2

Torque expression is given as,

$$T_m = \left(\frac{3}{\omega_s}\right) \cdot \frac{s E_2^2}{\left[R_2^2 + (s X_2)^2\right]} \cdot R_2$$

For large value of slip, $(s x_2)^2 \gg R_2^2$

At starting (s = 1), torque expression is given as,

$$T_{st} = \left(\frac{3}{\omega_s}\right) \cdot \frac{E_2^2}{X_2^2} \cdot R_2$$

$$T_{st} \, \alpha \, \frac{E_2^2 . R_2}{X_2^2}$$

Where,

α = Proportionality

Starting torque increases linearly with rotor resistance. If leakage reactance is high, starting torque is reduced. Torque is proportional to the square of the applied voltage.

Torque-slip Characteristics

The induced torque is zero at synchronous speed. The curve is nearly linear between no-load and full load. In this range, the rotor resistance is greater than the reactance where the rotor current and torque increases linearly with the slip. There is a maximum possible torque that cannot be exceeded. This torque is called as the breakdown torque and it is 2 to 3 times the rated full-load torque.

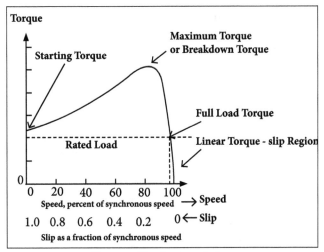

Torque-slip characteristics.

The starting torque of the motor is slightly higher than its full-load torque where the motor carries any load. The torque of the motor for a given slip varies as the square of the applied voltage. If the rotor is driven faster than the synchronous speed, it will run as a generator which converts mechanical power to electric power.

Full load torque is given as,

$$T_{Full \, load} < T_m$$

Condition for Maximum Torque

Maximum torque (also called breakdown torque) occurs when $sX_2 = R_2$.

Slip at maximum torque is given as,

$$S_{maxT} = R_2 / X_2$$

Maximum torque (T_{max}) is given as,

$$T_{max} = \left(\frac{3}{\omega_s}\right) \cdot \frac{E_2^2}{2X_2}$$

The value of maximum torque does not depend on the rotor resistance. Slip at which it occurs depends on the rotor resistance.

The below figure shows the complete torque-slip characteristics of motoring, generating and the braking region:

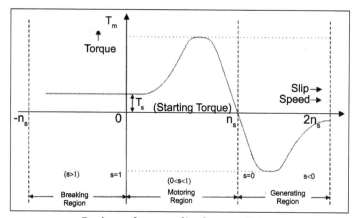

Regions of torque-slip characteristics.

Effect of Rotor Resistance on Torque-slip (Speed) Characteristics

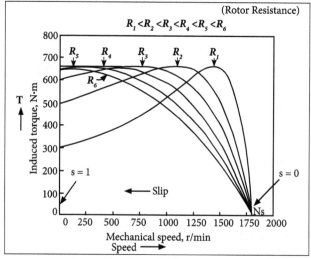

Torque-slip (speed) characteristics.

Efficiency

It is defined as the ratio of the output to that of the input. The induction motor receives power from the main supply through the stator windings. The input power is an electrical power. The power from the stator winding is transferred to the rotor via air gap by using induction principle. The output power is mechanical power which is obtained at the shaft of the motor to perform the mechanical work.

During this power transfer, some losses such as frictional losses, copper losses in rotor winding, copper losses in stator winding, core losses in stator winding and wind age losses takes place. Then the efficiency is determined by knowing the input and the output power.

Efficiency is defined as the ratio of the output to the input.

It can be expressed as,

$$\text{Efficiency}, \eta = \frac{\text{output}}{\text{input}}$$

Rotor efficiency of the three phase induction motor = Rotor output/Rotor input

= Gross mechanical power developed/Rotor input

$$= \frac{P_m}{P_2}$$

Efficiency of three phase induction motor = Power developed at shaft/Electrical input to the motor

Efficiency of the three phase induction motor is given as,

$$\eta = P_{out} / P_{in}$$

Single Phase Induction Motor

Basic Principle

When the stator of a single phase motor is fed with a single phase supply, it produces an alternating flux in the stator winding.

According to Faraday's law of electromagnetic induction, the alternating current flowing through the stator winding produces an induced current in the rotor bars. Such an induced current in the rotor will produce an alternating flux. Even after both the alternating fluxes are set up, the motor fails to start.

If an initial start is given to the rotor by an external force in either direction, then motor accelerates to its final speed and runs with its rated speed. Like any other electrical motor, asynchronous motor have two main parts namely rotor and stator.

Single Phase Induction Motor

Single phase induction motor is an AC motor where the electrical energy is converted to mechanical energy to perform some physical task. This induction motor requires only one power phase for their proper operation.

They are used in low power applications such as in domestic and industrial use. Simple construction, cheap cost, better reliability, easy to repair and better maintenance are some of its mark able advantages.

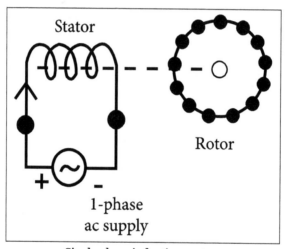

Single phase induction motor.

Construction of Single Phase Induction Motor

The main components of the single phase induction motor are stator and rotor. Stator is known as the stationary part. Single phase alternating supply is given to the stator winding. Rotor is the rotating part of the motor.

Rotor is connected to the mechanical load with the help of a shaft. A squirrel cage rotor is used here. It has a laminated iron core with many slots. Rotor slots are closed or semi-closed type.

The rotor windings are symmetrical and it is short circuited. An air gap exist between the rotor and the stator. They are used in refrigerators, clocks, drills, pumps, washing machines etc.,

The stator winding in the 1 phase induction motor has two parts such as main winding and auxiliary winding. Auxiliary winding is perpendicular to the main winding. In the single phase induction motor, the winding with more turns is known as main winding while the other wire is termed as auxiliary winding.

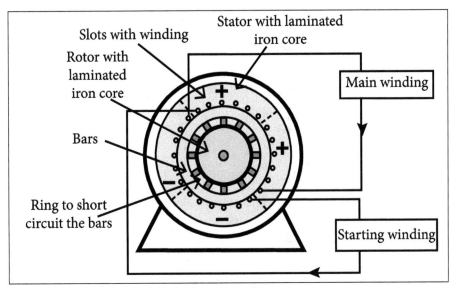

Single phase induction motor.

Types of Single Phase Induction Motors

Single-phase motors are generally built in the fractional-horse power range and it is classified as follows:

- Single-phase induction motor.

- AC series motor or universal motor.

- Repulsion motor:

    ○ Repulsion-start induction-run motor.

    ○ Repulsion-induction motor.

Single-phase Induction Motors

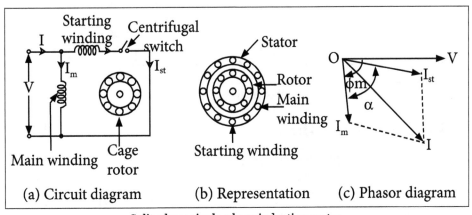

(a) Circuit diagram   (b) Representation   (c) Phasor diagram

Split-phase single-phase induction motor.

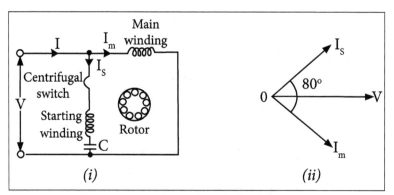

Capacitor type single-phase induction motor.

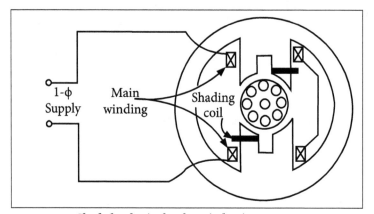

Shaded pole single-phase induction motor.

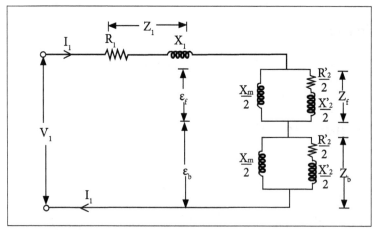

Equivalent circuit of single-phase induction motor.

Double Field Revolving Theory

According to the double field revolving theory, any alternating quantity can be resolved into two components. Magnitude of each component is equal to the half of the maximum magnitude of an alternating quantity. Both these component rotates in opposite direction to each other.

For example, a flux φ can be resolved into two components as $\phi_m/2$ and $-\phi_m/2$. Each of these components rotates in opposite direction i.e., If one $\phi_m/2$ is rotating in clockwise direction, then the other $\phi_m/2$ rotates in an anticlockwise direction.

When a single phase AC supply is given to the stator winding of single phase induction motor, it produces its flux of magnitude ϕ_m.

Let these two components of flux is referred as forward component of flux ϕ_f and backward component of flux ϕ_b. The resultant of these two component of flux at any instant of time gives the value of instantaneous stator flux at that particular instant.

$$\text{i.e.} \ \phi_r = \frac{\phi_m}{2} + \frac{\phi_m}{2}$$

$$\phi_r = \phi_f + \phi_b$$

At starting state, both the forward and backward components of flux are exactly opposite to each other and both of these components of flux are equal in magnitude. They cancel each other where the net torque experienced by the rotor at starting is zero. Hence, the single phase induction motors are not self-starting motors.

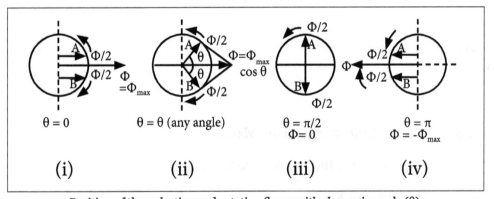

Position of the pulsating and rotating fluxes with change in angle (θ).

When the stator winding carries a sinusoidal current, a sinusoidal space distributed mmf whose peak or maximum value pulsates with time in the air gap. This sinusoidally varying flux (φ) is the sum of two rotating fluxes or fields where the magnitude of flux is equal to half the value of the alternating flux (φ/2). Both the fluxes rotate synchronously at the speed of n_s = 2f/p in opposite directions. This is shown in the above figure.

The above figure(i-iv) shows the resultant sum of the two rotating fluxes or fields as the time axis (angle) is changing from θ = 0° to π (180°). The flux or field rotating at synchronous speed in the anticlockwise direction i.e., the same direction as that of the motor is taken as positive induced emf (voltage) in the rotor conductors.

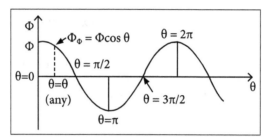

Pulsating (sinusoidal) flux as a function of space angle (θ).

Squirrel cage rotor with bars is short circuited through the end rings. The current flows in the rotor conductors and electromagnetic torque is produced in the same direction. The other part of flux or field which rotates at the same speed in the opposite (clockwise) direction is taken as negative.

Two torques are in the opposite direction and the resultant torque is the difference of the two torques. If the rotor is stationary, the slip due to forward (anticlockwise) rotating field is $s_f = 0.1$. The two torques are equal and opposite so that the resultant torque is 0.0 (zero). So, there is no starting torque in a single-phase induction motor where $S_b = 0.1$.

If the motor (rotor) is started or rotated in the anticlockwise (forward) direction, the forward torque is more than the backward torque where the resultant torque is positive. The motor accelerates in the forward direction where the forward torque is more than the backward torque.

The resultant torque is positive since the motor rotates in the forward direction. Speed of the motor depends on the load torque including the losses.

Starting Methods

Types of Single Phase Induction Motors

The method used to start the motor are as follows:

- Capacitor start motors.

- Capacitor-run motors.

- Capacitor-start and run motors.

- Shaded-pole motors.

Capacitor-start Motor

A larger capacitor is used to start a single phase induction motor via the auxiliary winding and it is switched out by a centrifugal switch once the motor is up to its speed. The auxiliary winding may have more turns of heavier wire than used in a resistance split-phase motor to mitigate the excessive temperature rise.

More starting torque is available for heavy loads like an air conditioning compressors. This motor configuration works well so that it is available in multi-horse power (multi-kilowatt) sizes as induction motor.

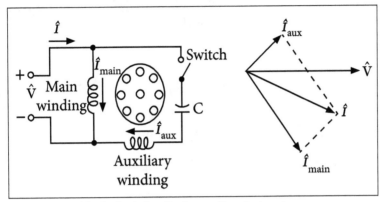

Equivalent circuit of capacitor start motors.

Types of single phase capacitor-start motor:

- Single voltage externally reversible.
- Single voltage non-reversible.

Applications

It is used in elevators, small pumps, compressors, blowers, refrigerators and fans.

Capacitor Run Motors

Capacitor is connected in series with the start winding to increase the running efficiency. They use run-capacitors that are designed for continuous duty that are energized for the entire time during the operation of the motor.

Working Principle of the Single Phase Capacitor-run Motors

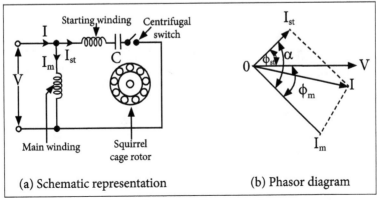

(a) Schematic representation

(b) Phasor diagram

Single phase capacitor-run motor.

In capacitor-run motors, a run-capacitor is connected to the start winding of the motor and it constantly energizes the start winding while the motor is running. This creates a 90° phase change between the start winding current and the run winding current to make a two phase motor. As a result, a rotating magnetic field is created within the motor which rotates the rotor more efficiently.

Advantages:

- The capacitor remains in the circuit at all times and no centrifugal switch is required.

- They are more efficient than other type of motors.

- They have low vibration and less noise under full load condition.

Disadvantages:

- It is used for applications which require low starting torque and high efficiency such as small compressors, pumps and fans.

- Since capacitor start motors that have low starting torque, they can't be used in applications with severe starting conditions.

Capacitor Start-and-run Motors

Capacitor-start-and-run motors or permanent-split capacitor motors are single phase induction motors where capacitors are connected in the circuit during both the starting and the running period.

In this type of motors, both the start winding and the run winding are permanently connected to the power source through a capacitor at all times.

Types of Single Phase Capacitor Start-and-run Motors:

(i) Single value capacitor start-and-run motors.

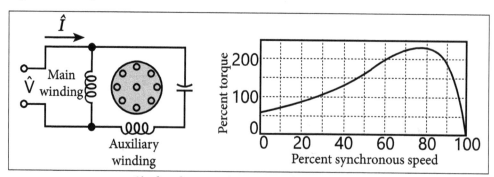

Single value capacitor start-and-run motors.

(ii) Two value capacitor start-and-run motors.

Two values of capacitance will be obtained using the following two different methods:

- By using two capacitors in parallel.

- By using a step up transformer.

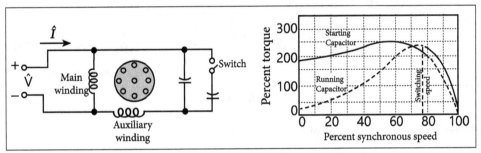

Two value capacitor start-and-run motors.

Advantages:

- It has the ability to start heavy loads.

- It has higher efficiency and power factor.

- It has the ability to develop 25% of overload capacity.

- It is used in extremely quiet operation.

Applications of single phase capacitor start-and-run motors:

- Single value capacitor start-and-run motors are used in applications that require low starting torque such as blowers, fans and voltage regulators.

- Two value capacitor start-and-run motors are used in applications that require variable speed such as fans, air handlers and blowers.

Shaded Pole Motors

A shaded pole motor is a single phase induction motor where one or more short circuited windings act only on a portion of the magnetic circuit. Winding is a closed copper ring which is embedded in the face of the pole called as the shaded pole which provides the required rotating field for starting purpose.

Working Principles of the Single Phase Shaded Pole Motors

When an alternating current is passed through the field winding or main winding surrounding the whole pole, the magnetic axis of the pole shifts from the unshaded part to the shaded part which is analogous with the actual physical movement of the pole. As a result, the rotor starts rotating in the direction of this shift from the unshaded part to the shaded part.

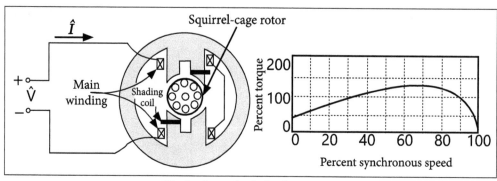

Single phase shaded pole motors.

Advantages of single phase shaded-pole motors:

- It has tough surface.

- It is reliable and cheap.

- Construction of single phase shaded-pole motor is simple.

Disadvantages of single phase shaded-pole motors:

- Overload capacity is low.

- Starting torque is low.

- Efficiency is low (5% for tiny sizes and 35% for higher ratings).

Applications of single phase shaded-pole motors:

- Due to its low starting torque, the shaded pole motor is used in small fans, toys, hair dryers and ventilators.

Electronic Devices and Circuits

4.1   Types of Materials

By considering their electrical properties, we can divide the materials into three groups as conductors, insulators and semiconductors.

Classification of Materials

Conductors	Materials with lots of free electrons are called as conductors. These electrons easily flow through the material. Example: All metals and semi-metals like carbon-graphite, antimony and arsenic.
Insulators	The materials that have a very few free electrons are called as insulators. Example: Plastic, glass and wood.
Semiconductors	These materials lie between the extremes of the good conductors and good insulators. They are crystalline materials which are insulators when pure but it will conduct when the impurity is added in response to light, heat, voltage etc., Example: Elements like silicon (Si), germanium (Ge), selenium (Se) and compounds like gallium arsenide (GaAs) and indium antimonide (InSb).

Electronics deals with the flow of electrons through vacuum, gas or semiconductors. Electronic devices like diodes, transistors, field effect transistors (FETs), silicon controlled rectifiers (SCRs), opto-electronic devices, and resistors, inductors, capacitors, etc. form circuits of electronic gadgets, equipment, and control systems.

An electronic device consists of integrated circuits which have several diodes, transistors, resistors, capacitors, etc. mounted on a single chip. Electronic components like diodes, transistors, SCRs, etc. are made of semiconductor materials.

As we know, all materials are classified into three categories, namely conductors, semiconductors, and insulators. Gold, silver, copper, aluminium, etc., are conducting materials. Conducting materials have a large number of free electrons in their atomic structure which allow flow of current. Rubber, ceramic, glass, wood, paper, bake lite, mica, etc. are insulating materials. In these materials no free electrons are available and as such no current should flow through them.

Substances like germanium, silicon, carbon, etc. are called semiconducting materials. Atoms of these materials binds themselves through sharing of electrons in their outermost shell or orbit. Such bonds are called covalent bonds. At absolute zero degree temperature, semiconducting materials behave like insulators as no free electrons are available for conduction. However, with increase of temperature or on application of voltage, some electrons become free electrons by breaking away from their covalent bonds and create a current flow. That is why these materials are called semiconductors.

4.1.1 N type and P Type Materials

There are two basic groups or classifications that can be used to define the different semiconductor types as follows:

- Intrinsic material.

- Extrinsic material.

Intrinsic Material: They are chemically very pure. It possesses a very low conductivity level with very few numbers of charge carriers namely holes and electrons in equal quantities.

Extrinsic Material: When a small amount of impurity is added to the basic intrinsic material, it is known as extrinsic material. This doping uses an element from a different periodic table group and it will either have more or less electrons in the valence band than the semiconductor itself. This creates either an excess or shortage of electrons. In this way, two types of semiconductor namely n-type and p-type are available where electrons are negatively charged carriers.

N-type Material

It has an excess of electrons. In this way, free electrons are available within the lattices and their overall movement is in one direction under the influence of a potential difference that results in an electric current. In an N-type semiconductor, the charge carriers are electrons.

P-type Semiconductor

In a p-type semiconductor material, there is a shortage of electrons i.e., there are holes in the crystal lattice. Electrons move from one empty position to another when the holes are moving.

It will happen under the influence of a potential difference and the holes flow in one direction resulting in an electric current. It is harder for holes to move than for free electrons to move and the mobility of holes is less than that of free electrons where holes are positively charged carriers.

Germanium

This type of semiconductor material was used in many early devices from radar detection diodes to the first transistors. Diodes show a higher reverse conductivity and transistors suffer from thermal runaway. It offers a better charge carrier mobility than silicon and it is used in some RF devices.

Silicon

Silicon is the most widely used semiconductor material. It is easy to fabricate and provides good general electrical and mechanical properties. When used for integrated circuits, it forms high quality silicon oxide which is used for insulation layers between different active elements of the integrated circuit.

N-type materials are formed by adding group 5 elements (pentavalent impurity atoms) to the semiconductor crystals and conduct the electric current by the movement of electrons.

N-type semiconductors:

- The impurity atoms are pentavalent elements.

- Impurity elements with solid crystal give a large number of free electrons.

- Pentavalent impurities are also called as donors.

- Doping gives the less number of holes in relation to the number of free electrons.

- Doping with group 5 elements results in positively charged donors and negatively charged free electrons.

- P-type materials are formed when group 3 elements (trivalent impurity atoms) are added to the solid crystal. In these semiconductors, the current flow is due to the holes.

P-type semiconductors:

- The impurity atoms are trivalent elements.

- Trivalent elements results in excess number of holes which always accepts electrons. Hence, trivalent impurities are called as acceptors.

- Doping gives the less number of free electrons in relation to the number of holes.

- Doping results in negatively charged acceptors and positively charged holes.

- Both p-type and n-type are electrically neutral on their own because the contribution of electrons and holes required for conducting electrical current are equal due to electron-hole pair.

- Both boron (B) and antimony (Sb) are called as metalloids because they are the most commonly used doping agents for the intrinsic semiconductor to improve the properties of conductivity.

4.2   PN Junction: Forward and Reverse Bias

As thin layers of P and N-type semiconductors are joined to form a junction, as shown in the below figure, a certain phenomenon takes place immediately:

- The majority holes form P-side diffuse into N-side and vice versa.

- Recombination of electrons and holes in a narrow region on both sides of the junction results in uncovered fixed positive ions on N-side and fixed negative ions on P-side.

- This is the depletion region where no free electrons and holes are present.

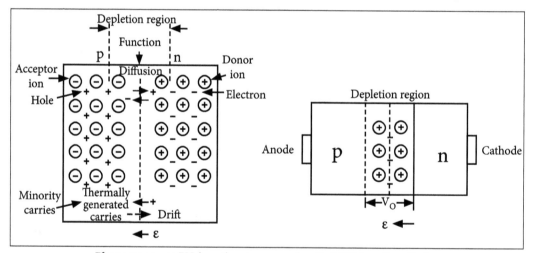

Phenomenon at PN-junction An open-circuited PN-junction diode.

- The electric field set up by the positive and negative ions prevents further flow of electrons and holes.

- The electric field causes the movement of minority carriers in opposite direction, a drift current.

- In steady-state, there is no net current flow across the junction.

- The simplified diagram of an open-circuit PN-junction diode is drawn in the below figure where V_0 = constant potential. The P-side terminal is called the anode and the N-side terminal is the cathode. The symbol of diode is shown in the below figure.

4.2.1 Forward and Reverse Bias

Reverse Biased PN Junction Diode

If the diode is connected in a reverse bias condition, a positive voltage is applied to the n-type material, whereas, a negative voltage is applied to the p-type material.

When the positive voltage is applied to the n-type material, it attracts electrons towards the positive electrode and away from the junction while the holes in the p-type end are attracted away from the junction towards the negative electrode.

The depletion layer grows wider because of a lack of electrons and holes and it presents a high impedance path such as an insulator. High potential barrier prevents the current from flowing via the semiconductor material.

Increase in the Depletion Layer Due to Reverse Bias

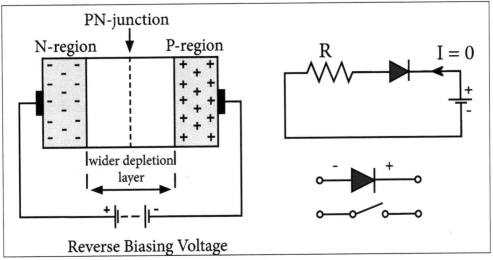

Reverse bias.

This condition represents a high resistance value to the PN junction and zero current flows through the junction diode with an increase in bias voltage. A very small leakage current flows through the junction which is measured in micro-amperes (μA).

When the reverse bias voltage V_r is applied to the diode, it increases to a sufficiently high enough value which will overheat the PN junction diode and fails because of the avalanche effect around the junction. This will affect the size of the diode and results in the flow of maximum circuit current.

Avalanche effect has practical applications in voltage stabilizing circuits where a series limiting resistor is used with the diode for limiting this reverse breakdown current to preset the maximum value and a fixed voltage output is produced across the diode. Such types of diodes are commonly known as zener diodes.

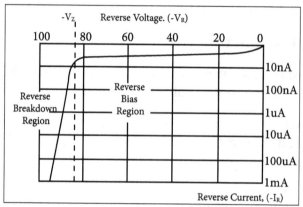

Reverse characteristics curve of a junction diode.

Forward Biased PN Junction Diode

If the diode is forward biased, a negative voltage is applied to the n-type material and a positive voltage is applied to the p-type material. When this external voltage is greater than the value of the potential barrier (0.7 volts for silicon and 0.3 volts for germanium), current starts to flow.

The negative voltage repels or pushes the electrons towards the junction by giving them the energy to cross over and combines with the holes that are being pushed in the opposite direction where the junction has positive voltage.

This results in a characteristics curve where zero current is flowing up to this voltage point known as the knee point on the static curves and then a high current flows through the diode with a little increase in the external voltage as shown in the below figure:

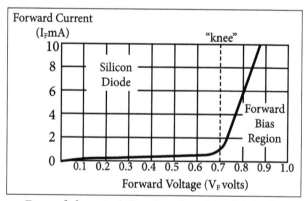

Forward characteristics curve for a junction diode.

The application of forward biasing voltage on the junction diode will result in the depletion layer which is very thin and narrow that represents a low impedance path through the junction thereby allowing high currents to flow. The point at which this sudden increase in current takes place is known as the knee point and it is represented on the static I-V characteristics curve.

Diode Relationship

$$I_D = I_S \left(e^{kV_D/T_k} - 1 \right) \qquad ...(1)$$

Where,

I_S = reverse saturation current

$k = 11{,}600/\eta$; $\eta = 1$ for Ge and $\eta = 2$ for Si for low current, below the knee of the curve and $\eta = 1$ for both Ge and Si for higher level of current beyond the knee (shown in the below figure).

$$T_k = T_c + 273°$$

And,

T_C = Operating temperature (25°C)

The plots of above equation (1) for Ge and Si diodes are drawn to scale in the below figure. The sharply rising part of the curve extended downward meets the VD axis, which is indicated as,

V_T = offset, threshold or firing potential.

It is quite accurate to assume that $I_D = 0$ up to V_T and then increases almost linearly at a sharp slope.

The values of VT are,

$V_T = 0.7$ V for Si diode

$V_T = 0.3$ V for Ge diode

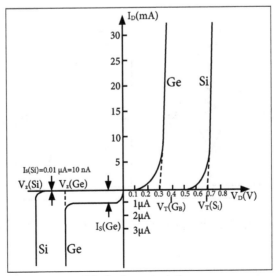

Diode characteristics.

Problems

1. An Si diode has I_s = 10 nA operating at 25°C. Let us calculate I_D for a forward bias of 0.6 V.

Solution:

We take $\eta = 2$

$$T_k = 25° + 273° = 298°$$

$$k = \frac{11,600}{2} = 5,800$$

$$kV_D / T_k = \frac{5800 \times 0.6}{298} = 11.68$$

$$e^{11.68} = 117930$$

Then,

$$I_D = 10(117930 - 1) = 50 \times 0.117929 \times 10^6 \, nA$$

$$= 0.586 \, mA, \text{ negligible.}$$

This justifies the choice of $\eta = 2$

2. A silicon diode is connected across a 3 V supply with a series resistance of 20 Q. Neglecting diode resistance, let us calculate the diode current.

Solution:

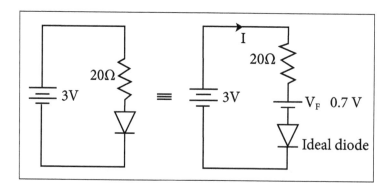

A silicon diode has V_F = 0.7 V. The diode equivalent circuit has been shown. Applying Kirchhoff's voltage law,

$$+3V - 20\, I - 0.7\, V = 0$$

$$20I = 3 - 0.7 = 2.3\,V$$

or,

$$I = \frac{2.3}{20} = 0.115A$$

4.3 Semiconductor Diodes

p- type and n-type Semiconductor

p- type semiconductor is formed by adding trivalent impurities to a pure or intrinsic semiconductor, whereas, the n-type semiconductor is formed by adding pentavalent impurities to the pure or intrinsic semiconductor.

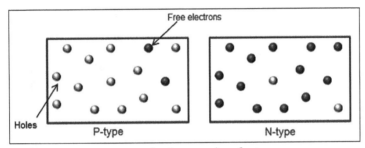

p- type and n-type semiconductor.

In p-type semiconductor, holes are the majority carriers, whereas, free electrons are the minority carriers. In n-type semiconductors, free electrons are the majority carriers, whereas, holes are the minority carriers.

Flow of Free Electrons and Holes

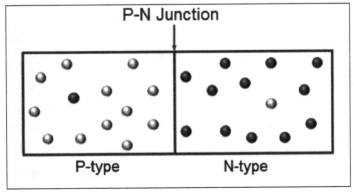

Flow of free electrons and holes.

If p-type semiconductor is joined with n-type semiconductor, then p-n junction is formed. The region in which the p-type and n-type semiconductors are joined is called as p-n junction. This p-n junction separates the n-type semiconductor from p-type semiconductor.

In n-type semiconductor, large number of free electrons are present which is repelled and they try to move from a region of high concentration to a region of low concentration. Near the junction, free electrons and holes are close to each other.

According to Coulomb's law, there exist a force of attraction between the opposite charges. Hence, the free electrons from n-side is attracted towards the holes at the p-side. Thus, the free electrons move from n-side to p-side and holes move from p-side to n-side.

Positive and Negative Charge at p-n Junction

The free electrons that are crossing the junction from the n-side provide extra electrons to the atoms on the p-side by filling holes in the p-side atoms. The atom that gains extra electron at p-side has more number of electrons than protons. When an atom gains an extra electron from the outside atom, it will become negative ion.

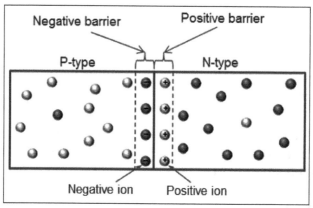

Positive and negative charge at p-n junction.

Each free electron crossing the junction from n-side to fill the hole in the p-side creates a negative ion at the p-side. Similarly, each free electron that left the parent atom at n-side to fill the hole in the p-side creates a positive ion at n-side.

Negative ion has more number of electrons than protons. Hence, it is negatively charged. Thus, a net negative charge is build at the p-side of the p-n junction. Similarly, positive ion has more number of protons than electrons. Hence, it is positively charged. Thus, a net positive charge is built at the n-side of the p-n junction.

The net negative charge at p-side of the p-n junction prevents the flow of free electrons from n-side to p-side since the negative charge present at the p-side of the p-n junction repels the free electrons. Similarly, the net positive charge at n-side of the p-n junction prevents the flow of holes from p-side to n-side.

Immobile positive charge at n-side and immobile negative charge at p-side near the junction acts like a wall or barrier that prevents the flow of free electrons and holes. The region near the junction where flow of charge carriers are decreased over a given time results in empty charge carriers or full of immobile charge carriers is known as depletion region.

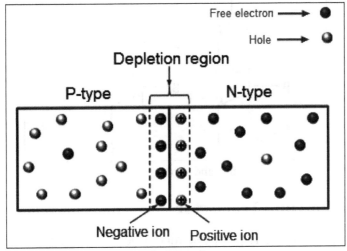

Depletion region.

The depletion region is also known as depletion layer, depletion zone, space charge region or space charge layer. The depletion region acts like a wall between the p-type and n-type semiconductor and it prevents the flow of free electrons and holes.

Knee Voltage of Diode

The voltage above which the diode operates in normal region is called as knee voltage of a diode.

4.4   Bipolar Junction Transistor

It is a three terminal semiconductor device consisting of two p-n junctions which can amplify or magnify a signal. It is a current controlled device.

The three terminals of the BJT are the collector, base and the emitter. A signal of small amplitude when applied to the base is available in the amplified form at collector of the transistor. This is the amplification provided by the BJT.

Its modes are as follows:

- Common base (CB) mode.

- Common emitter (CE) mode.

- Common collector (CC) mode.

Bipolar transistor has three region namely emitter, base and collector.

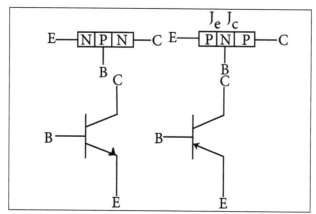

npn and pnp transistor.

n-p-n Bipolar Junction Transistor

In n-p-n bipolar transistor, one p-type semiconductor resides between the two n-type semiconductors. The diagram below shows a n-p-n transistor. Now I_E, I_C is the emitter current and collector current and V_{EB} and V_{CB} are emitter base voltage and collector base voltage.

According to the convention, if the emitter, base and collector current (I_E, I_B and I_C) current goes into the transistor, then sign of the current is taken as positive and if the current goes out from the transistor, then the sign is taken as negative.

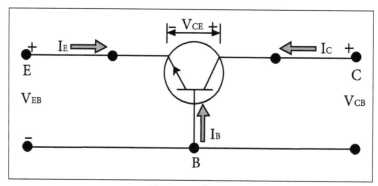

NPN transistor.

Different values of current and voltage inside the n-p-n transistor is shown in the below table:

Transistor type	I_E	I_B	I_C	V_{EB}	V_{CB}	V_{CE}
n - p - n	-	+	+	-	+	+

p-n-p Bipolar Junction Transistor

For p-n-p bipolar junction transistor, the n-type semi-conductors is sandwiched between two p-type semiconductors. The diagram of a p-n-p transistor is shown in the below figure:

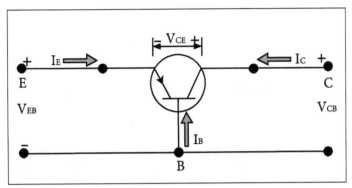

PNP transistor.

For p-n-p transistors, current enters the transistor through the emitter terminal. Like any BJT, the emitter-base junction is forward biased and the collector-base junction is reverse biased. We could tabulate the emitter, base and collector current as well as the emitter base, collector base and the collector emitter voltage for p-n-p transistors.

Transistor type	I_E	I_B	I_C	V_{EB}	V_{CB}	V_{CE}
p - n - p	+	-	-	+	-	-

Mechanism of p-n-p BJT

The basic operation of the transistor is described using the p-n-p transistor. The p-n junction of the transistor is forward-biased, whereas, the base-to-collector is without a bias as shown in the below figure:

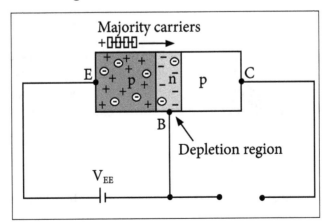

Forward-biased junction of a p-n-p transistor.

The middle portion is termed as the base (B) while the two end portions are known as the emitter (E) and collector(C). The junction between the emitter and the base is called as the emitter-base junction or the emitter junction (J_E). The junction between the collector and the base is called as the collector-base junction or the collector junction (J_C).

The depletion region reduces in width due to the applied bias which results in a heavy flow of majority carriers from p-type to the n-type material. The forward bias on the emitter-base junction will cause current to flow.

This flow of current consists of two components as follows:

- Holes are injected from emitter to base.

- Electrons are injected from base to emitter.

The design of the transistor ensures that the base region is fabricated very lightly compared to the emitter or collector regions. Holes are injected from the emitter to the base in large numbers and the injection of electrons from the base region is neglected. Let us remove the base-to-emitter bias of the p-n-p transistor as shown in the below figure:

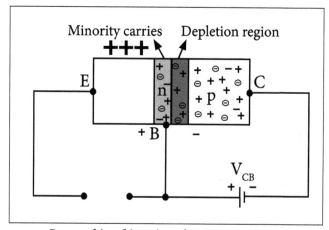

Reverse-biased junction of a p-n-p transistor.

The flow of majority carriers is zero which results in a minority-carrier flow. Thus, one p-n junction of a transistor is reverse-biased while the other is kept open. The operation of this device is easier when they are considered as separate blocks. The drift currents due to thermally generated minority carriers is neglected since they are very small.

4.4.1 Characteristics

To describe the behavior of any configuration, two characteristics are required:

- Driving point or input.

- Output.

Common Base Configuration

The circuit is shown in the below figure. Unless otherwise mentioned the transistor is NPN.

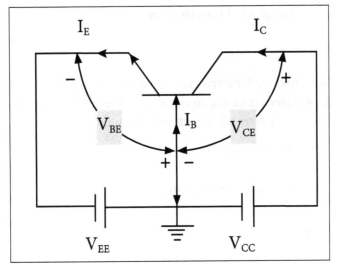

Common base configuration (NPN).

Input (Driving-Point) Characteristic (I_E vs V_{BE})

The EB junction is a forward-biased diode. A typical characteristic to scale is presented in the below figure. It is found that it is practically independent of V_{CB}. It can be approximated as a diode characteristic. Conduction begins for $V_{BE} = 0.7$ V as shown in the below figure (b).

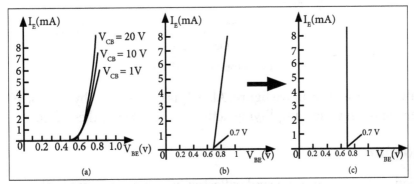

Driving point characteristic.

The current IE is controlled by adding a resistance in series with V_{EE}, while it will be assumed that,

$$V_{BE} = 0.7 \text{ V}$$

At any value of I_E, once the transistor starts conducting, ON state is shown in the above figure(c).

Output (Collector) Characteristics

These relate output current (I_C) with output voltage (V_{CB}) for varying values of input current (I_E). Typical output or collector characteristics are shown in the below figure.

The characteristics can be divided into three regions.

Active Region

The base-emitter junction is forward biased, while the collector-base junction is reverse biased. All the carriers that are injected into the emitter are swept away through the base to the collector. As a result, as already shown that,

$$I_C = \alpha I_E \text{ or } I_C = I_E \text{ as } \alpha = 1$$

α is the common-base forward current gain.

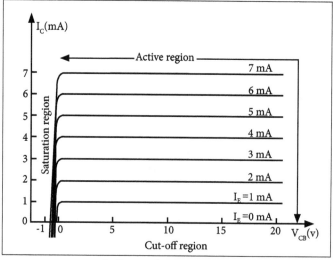

Collector characteristics.

This is easily seen from the above figure. Also, I_E is not affected by reverse bias V_{CB}. This is the linear region in which amplifying action of the circuit takes place.

Cut-off Region

In this region, both base-emitter and collector-base junctions are both reverse biased. As a result, $I_E = 0$ and so $I_C = 0$.

Saturation Region

In this region, both base-emitter and collector-base junctions are forward biased. This region is to the left of $V_{CB} = 0$. In this region, the collector current rises exponentially to the I_E value set by V_{BE} circuit as V_{CB} increases towards reverse bias.

Amplifying Action

Current amplification ≈ 1. The input current is transferred by the transistor to output.

Voltage amplification will result from low driving-point resistance (10 - 100 Ω) (shown in the below figure (b) and high output resistance.

The configuration is not used for amplification but serves certain special purposes.

Common-Emitter (CE) Configuration

It is the most frequently used configuration. Its circuit is shown in the below figure:

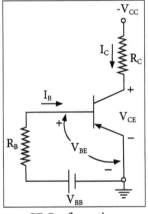

CE Configuration.

Input (Base) Characteristics

I_B vs V_{BE} for varying V_{CE} is shown in the below figure (b).

Output (Collector) Characteristics

I_C vs V_{CE} for various values of IB is shown in the below figure (a):

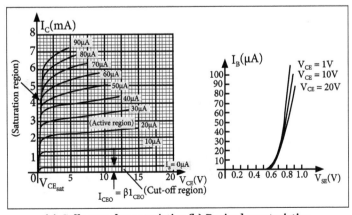

(a) Collector characteristics (b) Basic characteristics.

It is seen from the base characteristics that I_B is practically independent of V_{BE}. The base-emitter junction goes 'on' at $V_{BE} = 0.7$ V and then stays there, while I_B is adjusted by the external resistance R_B.

Active Region

Base-emitter is forward biased and collector-base is reverse biased. The collector characteristics are shown in the above figure (a). It is seen that IB is in μA and I_C in mA. These are related by β. The middle of this region is linear w.r.t. I_B and V_{CE}.

Cut-off Region

It is below $I_B = 0$; the EB junction becomes reverse biased but the corresponding $I_C \neq 0$.

Wkt, $I_C = \beta I_B + (1+\beta) I_{CBO}$

Or,

$$I_C = (1+\beta) I_{CBO} \approx \beta I_{CBO}; \text{ for } I_B = 0$$

$$\beta I_{CBO} = I_{CEO}$$

If $I_{CBO} = 1\mu A$, resulting from $I_B = 0$, then

$$I_{CEO} = 250 \times 1 \times 10^{-2} = 0.25\,\text{mA at} \beta = 250$$

Saturation Region

It is to the left of $V_{CE(sat)} = 0.2$ V. In this region, the CB junction becomes forward biased and I_B no longer controls I_C.

β_{dc} and β_{ac} lie in the middle region of the above figure (a), at any point corresponding to a certain V_{CE} say 10 V.

$$\beta = \beta_{dc} = \frac{I_C}{I_B}, \text{large signal gain}$$

For variation about this point along $V_{CE} = 10$ V,

$$\beta_{ac} = \frac{\Delta I_C}{\Delta I_B}$$

Study on typical collector characteristics shows that,

$$\beta_{ac} \approx \beta_{dc}$$

Relationship between DC Currents and Gains

$I_E = I_B + I_C$ $I_C = I_E - I_B$ $I_B = I_E - I_C$	$\alpha = I_C / I_E = \beta / (1+\beta)$ $\beta = I_C / I_B = \alpha / (1-\alpha)$
$I_B = I_C / \beta = I_E / (1+\beta) = I_E (1-\alpha)$	
$I_C = \beta I_B = \alpha I_E$	$I_E = I_C / \alpha = I_B (1+\beta)$

CC Configuration

Common collector configuration circuit is shown in the below figure. The collector is grounded and it is used as a common terminal for both the input and output. It is also called as grounded collector configuration. Base is used as an input terminal, whereas, emitter is the output terminal.

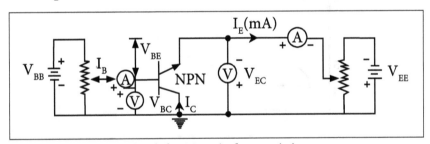

Circuit for CC static characteristics.

Input Characteristics

It is defined as the characteristic curve drawn between input voltage to input current, whereas, output voltage is constant.

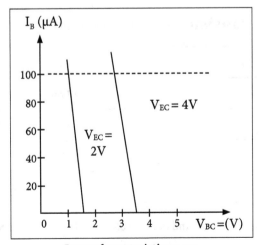

Input characteristic curve.

To determine input characteristics, the emitter base voltage V_{EC} is kept at a fixed value. The base collector voltage (V_{BC}) is increased in equal steps and the corresponding increase in IB is noted. This is repeated for higher fixed values of V_{CE}.

A curve is drawn between base current and the base collector voltage at constant collector emitter voltage as shown in the above figure.

Output Characteristics

It is defined as the characteristic curve drawn between output voltage to output current, whereas, input current is constant. To determine the output characteristics, the base current I_B is kept constant at zero and emitter current IE is increased from zero by increasing V_{EC}. This is repeated for higher fixed values of I_B. From the output characteristic, it is seen that, for a constant value of I_B, I_E is independent of V_{EB} and the curves are parallel to the V_{EC} axis.

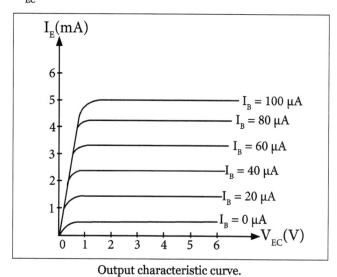

Output characteristic curve.

Performance of a Transistor in Various Configurations

Property	CB	CE	CC
Input resistance	Low	Moderate	High
Output resistance	High	Moderate	Low
Voltage gain	High	High	≈1
Current gain	≈1	High	High
Phase shift between input voltage and output voltage	0° to 360°	180°	0° or 360°

4.5 Field Effect Transistors and Transistor Biasing

FET is a three terminal (i.e., drain, gate and source) unipolar voltage controlled device. There are two types of such devices namely MOSFET (Metal oxide semiconductor field effect transistor) and JFET (Junction field effect transistor).

Junction gate field effect transistor (JFET or JUGFET) is the simplest type of field effect transistor which is used as electronically controlled switches, amplifiers or voltage controlled resistors.

JFET is classified into n-channel JFET, where current conduction occurs due to electrons and p-channel JFET, where current conduction occurs due to holes. In JFET, gate to source junction is always in reverse bias condition and drain is always at high potential than the source.

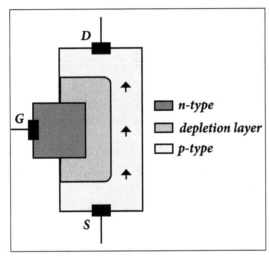

Structure of JFET.

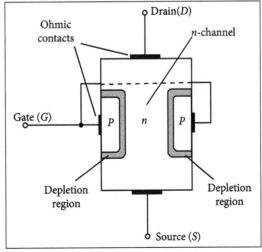

Junction field effect transistor (JFET).

JFET is a long channel of semiconductor material which contains an abundance of positive charge carriers or holes (p-type) or negative charge carriers or electrons (n-type). Ohmic contacts at each end form the source (S) and drain (D).

Unlike bipolar transistors, JFET's are exclusively voltage controlled and they do not need a biasing current. Electric charge flows through a semiconducting channel between the source and drain terminals. By applying a reverse bias voltage to the gate terminal, the channel is pinched and the electric current is impeded or switched off completely.

JFET's have an n-type or p-type channel. In n-type, if the voltage applied to the gate is less than that applied to the source, and then the current will be reduced. Electric current from source to drain in a p-channel JFET is restricted when a voltage is applied to the gate.

PN junction is formed on one side of the channel or at both sides of the channel or surrounding it using a region with doping which is opposite to that of the channel and biased using an ohmic gate contact (G).

JFET is usually ON when there is no potential difference between its source and gate terminals. If a potential difference of proper polarity is applied between its gate and source terminals, JFET will be more resistive to current flow, which means that, less current will flow in the channel between the source and drain terminals. Thus, JFET's are also known as depletion mode devices.

JFET has a large input impedance which means that it has a negligible effect on external components or circuits connected to its gate.

Schematic Symbols

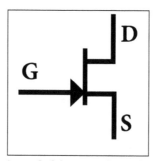

Circuit symbol for an n-channel JFET.

The JFET gate is drawn at the middle of the channel. This symmetry suggests that source and drain are interchangeable and the symbol should be used only for those JFET's where they are indeed interchangeable.

In every case, the arrow head shows the polarity of the PN junction formed between the channel and the gate. In ordinary diode, the arrow points from P to N which gives the direction of conventional current when forward biased.

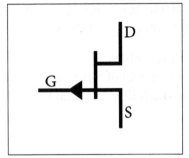

Circuit symbol for a p-channel JFET.

Construction

Depending on the structure, the junction field effect transition (JFET's) is divided into the following two categories:

- N-channel JFET.

- P-channel JFET.

The basic construction of an N-channel JFET is shown in below figure (a). It consists of an N-type semiconductor bar with two P-type heavily doped regions diffused on opposite sides of its middle part. The P-type regions form two PN junctions. The space between the junctions is called as channel. Both the P-type regions are connected internally and single wire is taken out in the form of a terminal called the gate (G).

The electrical connection made at both ends of the N-type semiconductor are taken out in the form of two terminals called as drain (D) and source(S).

Drain (D) is a terminal through which electrons leave the semiconductor bar and source(S) is a terminal through which the electrons enter the semiconductor.

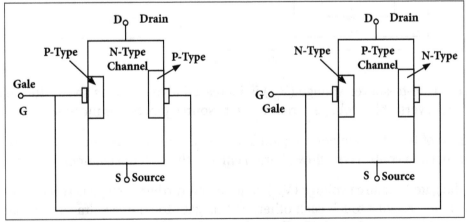

(a) n-channel JFET (b) p-channel JFET.

Whenever a voltage is applied across the drain and sources terminals, a current flows through the n-channel. The current consists of only one type of carriers. Therefore, the field-effect transistor is called as an unipolar device.

A p-channel JFET is shown in the above figure. Its construction is similar to that of N-channel JFET except that it consists of a p-channel and n-type regions. The current carriers in JFET are the holes which flow through the p-type channel.

Working Principle

Application of the negative gate voltage with respect to source reverse biases the gate source junction of an N-channel JFET. The effect of the reverse bias voltage forms depletion region within the channel.

When a voltage is applied between the drain and source with the DC supply voltage V_{DD}, the electrons flow from source to drain through the narrow channel which exist between the depletion region. This constitutes the drain current ID and its conventional direction is indicated from drain-to-source.

The value of drain current is maximum when no external voltage is applied between the gate and source and it is designated by the symbol I_{DSS}.

Operation of JFET

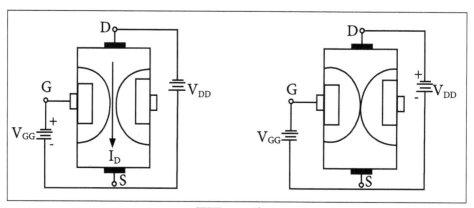

JFET operation.

When the gate-to-source voltage (V_{GS}) is increased above zero as shown in the above figure, the reverse-bias voltage across the gate source junction is increased.

As a result of this, the depletion region is widened. This reduces the effective width of the channel and controls the flow of drain current through the channel.

When the gate-to-source voltage (V_{GS}) is increased further, a stage is reached at which two depletion regions touch each other. At this gate-to-source voltage, the channel is completely blocked (or) pinched off and drain current is reduced to zero.

The gate-to-source voltage (V_{GS}) at which the drain current is zero is called as pinch off voltage. It is designated by the symbol V_p (or) V_{GS} (off). The value of pinch off voltage V_p is negative for N-channel JFET.

Comparison of n-channel JFET and p-channel JFET

S. No	n-channel JFET	p-channel JFET
1.	In n-channel, the current carriers are electrons.	In p-channel, current carriers are holes.
2.	Mobility of electrons is almost twice that of holes in p-channel.	Mobility of holes is poor.
3.	Input noise is low.	Input noise is high.
4.	Trans-conductance is high.	Trans-conductance is low.

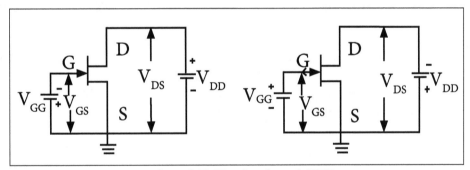

n- channel JFET and p-channel JFET.

The BJT is a current-controlled device while the JFET transistor is a voltage-controlled device i.e., the output current is controlled by an electric field created by the applied devices.

Pinch off Voltage

The drain to source voltage (V_{DS}) at which the channel pinch off occurs and drain current remains constant to its maximum value is called as pinch-off voltage.

Amplification Factor in JFET (β)

The ratio of change in collector current (ΔI_C) to the change in base current (ΔI_B) is known as amplification factor i.e.,

$$\beta = \frac{\Delta I_C}{\Delta I_B}$$

4.5.1 Transistor Biasing

Bias design is used to obtain a desired drain quiescent current IDQ which does not vary too much with the wide spread in device parameters like IDSS and VP. In the data sheet, the minimum and maximum values for IDSS are given and they are spread over a range of between about four and ten to one. The minimum and maximum values for VP are also provided and they vary over a range of between about three and seven to one.

There is a strong correlation between the minimum I_{DSS} and the minimum V_P. These values are referred as I_{DSSmin}, V_{Pmin}, I_{DSSmax}, V_{Pmax}.

There are two circuit methods to accomplish the desired bias. In the first method as shown in the below figure (left), the source is connected to ground and a gate bias voltage is applied from V_{GG} through a series resistor R_G. This method is very sensitive to the spread in device parameters.

In the second method as shown in the below figure (right), a resistor R_S is connected in series with the source terminal and connected to a supply (V_{SS}). The gate bias voltage is applied from V_{GG} through a series resistor R_G.

This method has more flexibility and it minimizes the sensitivity to the spread of device parameters. There are two options. One is to set V_{GG} to a particular voltage and the required value for R_S is calculated. In the second method, R_S is set to a particular resistance and the required value for V_{GG} is calculated.

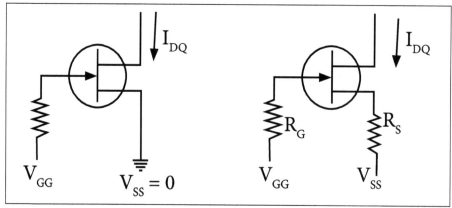

JFET bias circuit.

Three Basic Circuits

Three basic common-source circuits that can be used to establish a FET's operating point (Q-point) are as follows:

- Constant-voltage bias is used in RF and video amplifiers to employ small dc drain resistors.

- Constant-current bias is suitable for low-drift dc amplifier applications such as source followers and source-coupled differential pairs.

- Self-bias (also called source bias or automatic bias) is an universal scheme which is particularly valuable for ac amplifiers.

The Q-point established by the intersection of the load line and the $V_{GS} = -0.4V$ output characteristic of the below figure (a) provides a convenient starting point for the circuit comparison. The load line shows that a drain supply voltage V_{DD} of 30 V and a drain resistance RD of 39 kΩ are used.

The quiescent drain-to-source voltage V_{DSQ} is 16 V which allows large signal excursions at the drain. Maximum input signal variations of $\pm$ 0.2 V will produce output voltage swings of $\pm$ 7.0 V and a voltage gain of 35 where,

$$A_V = \frac{g_{fs} R_D}{1 + R_D g_{os}} \qquad ...(1)$$

g_{os} = Output conductance of JFET

Since,

$R_D g_{os}$ is negligible, we have,

$$A_V \approx g_{fs} R_D \qquad ...(2)$$

Constant-voltage Bias

Constant-voltage bias circuit is shown in the below figure (b) which is analyzed by superimposing a line for V_{GG} (Constant) on transfer characteristic of the FET.

The transfer characteristic is a plot of I_D vs V_{GS} for constant V_{DS}. Since the curve does not change with changes in V_{DS}, it is used to establish operating bias points.

When a bias load line is superimposed, allowable signal excursions become evident and input voltage, gate-source signal voltage and output signal current calculations are made graphically.

The heavy vertical line at $V_{GS} = -0.4$ V establishes the Q-point of figure (a). No voltage is dropped across the resistor R_G because the gate current is zero. RG isolates the input signal from the V_{GG} supply.

Excursions of the input signal (e_g) is in series with V_{GS} so that they are added algebraically to the fixed value of -0.4 V. The effect of signal variation shifts the bias line horizontally without change in its slope. The shifting bias line develops an output signal current as shown in the below figure (b).

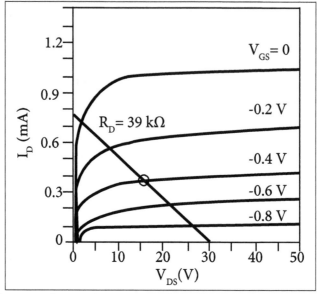

(a) Output characteristic curve - A large dynamic range is provided by the operating point at $V_{DSQ} = 15$ V, $I_{DQ} = 0.4$ mA and $V_{GSQ} = -0.4$ V.

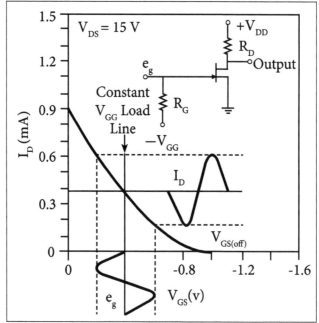

(b) Transfer curve - Constant-voltage bias is maintained by the V_{GG} supply as shown on this typical transfer curve. Input signal e_g moves the load line horizontally.

Constant-current Bias

The constant-current bias approach is shown in the below figure(c) for establishing the Q-point of the above figure (a) which requires 0.4 mA current source. For an ideal constant-current generator, the input signal excursions shift the bias line horizontally and gate-source voltage excursion is not produced.

This bias technique is limited to source followers; source coupled differential amplifiers and ac amplifiers where the source terminal is bypassed to the ground at the signal frequency.

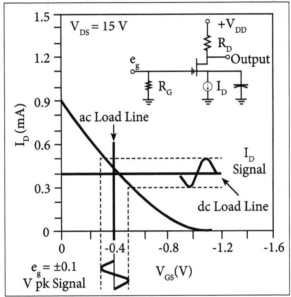

(c) Constant-current bias fixes the output voltage for any R_D where input signals do not affect the output unless the current source is bypassed.

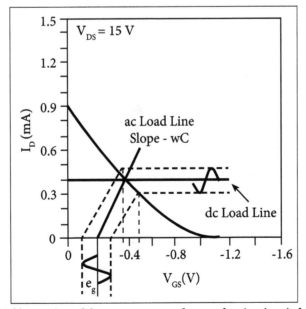

(d) Partial bypassing of the current source lowers the circuit gain by tilting the ac load line from the vertical where the capacitor drop subtracts from e_g.

An ac ground is provided by a bypass capacitor across the current source where a vertical ac bias line is established. The input signal variations will then translate the ac bias line horizontally and the signal development will proceed with constant-voltage biasing as shown in the above figure (c).

If the bypass capacitor does not provide a low reactance at the signal frequency, then the ac bias line will not be vertical. It will intersect the transfer curve at the Q-point but where a slope is equal to $-(1/X_C) = -C$ (above figure (d)). This will lower the gain of the amplifier because of the signal degeneration at the source. The input signal e_g is reduced by the drop across the capacitor which is given by the below equation (3).

$$V_{GS} = e_g - V_S = e_g - i_S X_C \qquad \dots(3)$$

From the above figure (d), the input signal shifts the operating point only by an amount equal to VGS which is the effective input signal. Since the signal frequency is decreases, the slope of the ac bias line decreases where the effective input signal is zero.

Self-bias (No Extra Supply)

The self-bias circuit shown in the below figure (e) establishes the Q-point by applying the voltage drop across source resistor RS to the gate. Since no voltage is dropped across R_S when $I_D = 0$, the self-bias load line passes through origin. Its slope is given as $-1/R_S = I_{DQ}/V_{GSQ}$.

The ac gain of the circuit is increased by shunting RS with a bypass capacitor as in constant-current case. The ac load line then passes through the Q-point with a slope $-(1/Z_S) = -(\omega_C + 1/R_S)$.

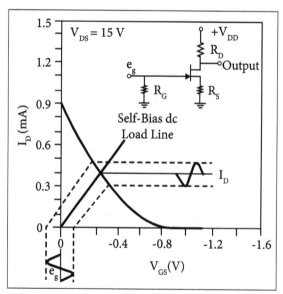

(e) Self-bias load line passes through the origin with a slope -1/R_S where bypassing R_S will steepen the slope and increases the gain of the circuit.

The circuit is biased automatically at the desired Q-point which do not require any extra power supply and provides a degree of current stabilization which is not possible with constant-voltage biasing.

Self-biasing

The fourth biasing method combining the advantages of constant-current biasing and self-biasing is obtained by combining the constant-voltage circuit with the self-bias circuit as shown in the below figure (f). A principal advantage of the configuration is that an approximation may be made constant-current bias without any additional power supply.

The bias load line is drawn through the selected Q-point and any desired slope is given by properly choosing V_{GG}. If V_{GG} is larger, then the RS will be larger.

All three circuits in the below figure (f) are equivalent. In the below figure, circuit (a) requires an extra power supply. The need for an additional supply is avoided in the below figure (b) since V_{GG} is derived from the drain supply. R_1 and R_2 are simply a voltage divider. To maintain the high input impedance of the FET, R_1 and R_2 must be very large.

Very large resistors are needed to derive the desired V_{GG} in every circuit application. In the below figure, circuit(c) overcomes this problem by placing a large R_G between the center point of the divider and the gate where R_1 and R_2 is small without lowering the input impedance.

When V_{GG} is increased, V_S increases and V_{DS} decreases. Therefore, with low V_{DD}, there may be a significant decrease in the allowable output voltage swing.

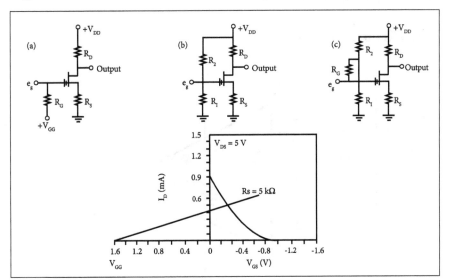

(f) All three combination-bias circuits are equivalent. They add constant-voltage biasing to the self-bias circuit to establish a reasonably flat load line without sacrificing dynamic range.

4.6   Operational, Inverting and Non-inverting Amplifier

Operational amplifiers are highly stable, high gain DC difference amplifiers. Since there is no capacitive coupling between their various amplifying stages, they can handle signals from zero frequency up to a few hundred kHz. They are used to perform mathematical operations on their input signal.

There are two inputs namely the inverting input (-) and the non-inverting input (+). These symbols have nothing to do with the polarity of the applied input signals.

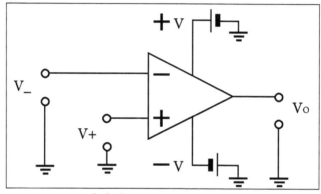

Symbol of an operational amplifier.

Connections to power supplies are shown in the above figure.

Characteristics of an Ideal op-amp

It has zero output impedance, zero input offset voltage, infinite input impedance, infinite voltage gain and infinite bandwidth.

Equivalent Circuit of an op-amp

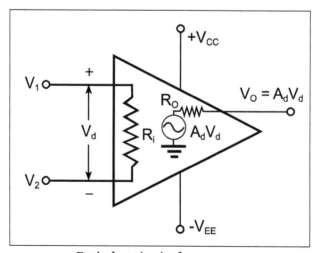

Equivalent circuit of an op-amp.

The two input voltages are V_1 and V_2. R_i is the input impedance of an op-amp. $A_d V_d$ is an equivalent Thevenin's voltage source and RO is the Thevenin's equivalent impedance of an op-amp.

This equivalent circuit is useful in analyzing the basic operating principles of an op-amp.

$$V_O = A_d \left(V_1 - V_2 \right) = A_d V_d$$

The output voltage V_O is directly proportional to the algebric difference between the two input voltages. An op-amp amplifies the difference between the two input voltages. It does not amplify the input voltages themselves. The polarity of the output voltage depends on the polarity of the difference voltage V_d.

Transfer Curve of an Ideal Voltage

The graphical representation of the output equation is shown in the below figure in which the output voltage V_o is plotted against differential input voltage V_d where gain A_d is constant.

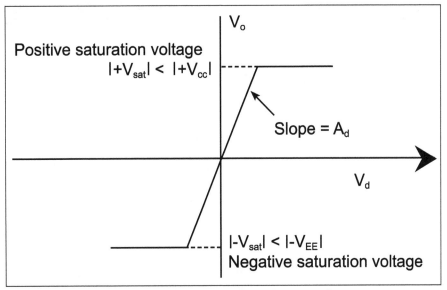

Graphical representation of an output equation.

The output voltage cannot exceed the positive and negative saturation voltages. These saturation voltages are specified for given values of supply voltages. Hence, the output voltage is directly proportional to the input difference voltage until it reaches the saturation voltages and the output voltage remains constant.

Hence, the curve is called as an ideal voltage transfer curve which is ideal because output offset voltage is zero. If the curve is drawn to scale, the curve is almost vertical because of the large values in A_d.

4.6.1 Inverting Amplifier and Non Inverting Amplifier

Inverting Amplifier

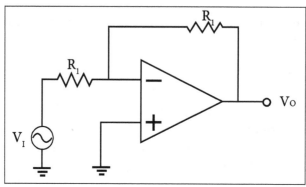

Inverting amplifier.

The operational amplifier is connected with the feedback to produce a closed loop operation. If the junction of an input and feedback signal at the same potential is at zero volts or ground, then the junction is termed as virtual earth.

Because of this virtual earth node, the input resistance of the amplifier is equal to the value of the input resistor R_{in} and the closed loop gain of an inverting amplifier is the ratio of the two external resistors.

Properties of an inverting amplifiers or any operational amplifier is as follows:

- There is no current at the input terminals.

- The differential input voltage is zero as $V_1 = V_2 = 0$.

By using these two rules, we can derive the equation for calculating the closed-loop gain of an inverting amplifier using first principles.

Current (i) flows through the resistor network is given as,

$$i = \frac{V_{in} - V_{out}}{R_{in} + R_f}$$

Therefore,

$$i = \frac{V_{in} - V_2}{R_{in}} = \frac{V_2 - V_{out}}{R_f}$$

Therefore,

$$i = \frac{V_{in}}{R_{in}} - \frac{V_2}{R_{in}} = \frac{V_2}{R_f} - \frac{V_{out}}{R_f}$$

$$\frac{V_{in}}{R_{in}} = V_2 \left[\frac{1}{R_{in}} + \frac{1}{R_f} \right] - \frac{V_{out}}{R_f}$$

And as,

$$i = \frac{V_{in} - 0}{R_{in}} = \frac{0 - V_{out}}{R_f} \qquad \frac{R_f}{R_{in}} = \frac{0 - V_{out}}{V_{in} - 0}$$

The closed Loop Gain (Av) is given as,

$$\frac{V_{out}}{V_{in}} = -\frac{R_f}{R_{in}}$$

Then, the closed-loop voltage gain of an inverting amplifier is represented as,

$$\text{Gain}\left(\text{Av}\right) = \frac{V_{out}}{V_{in}} = -\frac{R_f}{R_{in}}$$

$$V_{out} = -\frac{R_f}{R_{in}} \times V_{in}$$

The negative sign indicates that the inversion of an output signal with respect to the input is 180° out of phase and the feedback is negative.

The equation for the output voltage V_{out} shows that the circuit is linear in nature for a fixed amplifier gain as $V_{out} = Vi_n$ x Gain. This property is used for converting a smaller sensor signal to a much larger voltage.

Another useful application of an inverting amplifier is trans-resistance amplifier circuit.

Non Inverting Amplifier

In the non-inverting circuit configuration, input impedance R_{in} is increased to infinity and its feedback impedance R_f is reduced to zero.

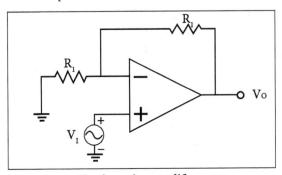

Non inverting amplifiers.

The output is connected directly back to the negative inverting input where the feedback is 100% and V_{in} is exactly equal to Vout where a fixed gain is unity. Since the input voltage V_{in} is applied to the non-inverting input, the gain of the amplifier is given as,

$$V_{out} = A\ V_{in}$$

$$V_{in} = V^+$$

$$V_{out} = V^-$$

Gain is given as,

$$A_v = \frac{V_{out}}{V_{in}} = +1$$

4.7 DAC and ADC

Digital to Analog Converter (DAC) is a device that transforms digital data into an analog signal. According to the NY Quist-Shannon sampling theorem, any sampled data can be reconstructed perfectly with bandwidth and NY Quist criteria.

A DAC can reconstruct sampled data into an analog signal with precision. The digital data may be produced from a microprocessor, Application Specific Integrated Circuit (ASIC), or Field Programmable Gate Array (FPGA), but ultimately the data requires the conversion to an analog signal in order to interact with the real world.

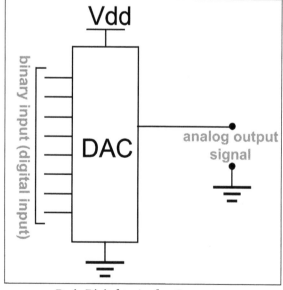

Basic Digital to Analog Converter.

DAC is classified as three types as follows:

- Weighted resistor DAC.

- R-2R and inverted R-2R DAC.

- Monolithic DAC.

Weighted Resistor DAC

One of the simplest circuit shown in the below figure uses a summing amplifier with a binary weighted resistor network.

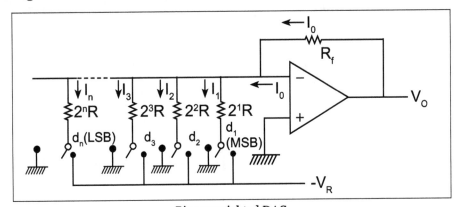

Binary weighted DAC.

It has n-electric switches d_1, d_2.... d_n controlled by a binary input word. Such switches are single pole double throw (SPDT) type.

If the binary output to a particular switch is 1, it connects the resistance to the reference voltage (-V_R). If the input bit is 0, the switch connects the resistor to the ground. In the above figure, output current I_o for an ideal op-amp is written as,

$$I_o = I_1 + I_2 + I_3 + ... + I_n$$

$$= \frac{V_R}{2R} d_1 + \frac{V_R}{2^2 R} d_2 + ... + \frac{V_R}{2^n R} d_n$$

$$I_o = \frac{V_R}{R} \left[d_1 2^{-1} + d_2 2^{-2} + ... + d_n 2^{-n} \right]$$

Output voltage is given as,

$$V_o = I_o R_f$$

$$V_o \frac{V_R}{R} R_f \left[d_1 2^{-1} + d_2 2^{-2} + ... + d_n 2^{-n} \right]$$

If $R_f = R$ and $K = 1$, we have,

$$V_{Fs} = V_R$$

Analog output voltage is a positive staircase as shown in the below figure for a 3-bit weighted resistor.

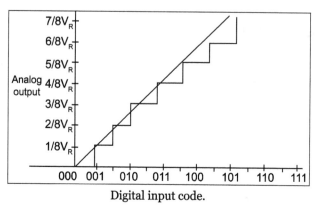

Digital input code.

Transfer characteristics of a 3-bit DAC:

- Although the op-amp is connected in inverting mode, it can also be connected in non-inverting mode.

- An op-amp works as a current to voltage converter.

- The polarity of the reference voltage is chosen in accordance with the type of the switch used. For TTL compatible switches, the reference voltage is +5V and output will be negative.

The accuracy and stability of a DAC depends on the accuracy of the resistors and tracking of each other with temperature.

Disadvantage:

- Wide range of resistor values are required. For better resolution, the input binary word length should be increased. When the number of bit increases, the range of resistance value increases.

- For a 12-bit DAC, the largest resistance required is 5.12 mΩ. If the smallest is 2.5 kΩ, then the fabrication of such a large resistance in IC is not practical.

- The voltage drop across a large resistor due to the bias current will affect the accuracy. For smaller values of resistors, the loading effect will occur.

- Finite resistance of the switches disturbs the binary-weighted relationship among the various currents particularly in the most significant bit positions where the current setting resistances are smaller.

Current Mode DAC

As the name implies, current mode DAC's operates based on the ladder currents. The ladder is formed by the resistance R in the series path and the resistance 2R in shunt path. Therefore, the current is divided into i_1, i_2, i_3,.... in in each arm. These currents are either diverted to ground bus (i_0) or to virtual-ground bus ($\bar{i}_0$).

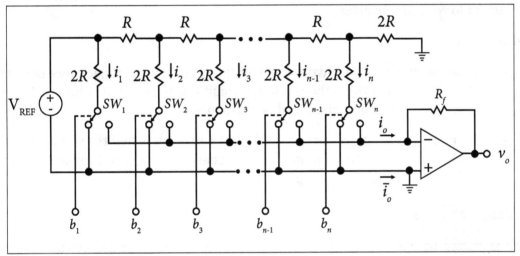

Current mode DAC.

Currents are given as follows,

$$i_1 = V_{REF} / 2R = \left(V_{REF} / R\right)_2^{-1}$$

$$i_2 = \left(V_{REF}\right)/2\big)/2R = \left(V_{REF} / R\right)_2^{-2}$$

....

$$i_n = \left(V_{REF} / R\right)_2^{-n}$$

Relationship between the currents are given as follows:

$$i_2 = i_1 / 2$$

$$i_3 = i_1 / 4$$

$$i_4 = i_1 / 8$$

$$i_n = i_1 / 2^{n-1}$$

$$V_0 = -R_f i_0$$

Gives,

$$V_o = -(R_f / R) V_{REF} \left(b_1 2^{-1} + b_2 2^{-2} + ... + b_n 2^{-n} \right)$$

The two currents i_o and $\bar{i}$ are complementary to each other and the potential of i_o bus is close to that of $\bar{i}$ bus. Otherwise, linearity errors will occur. The final op-amp is used as current to voltage converter.

Advantages:

- In the current mode or inverted ladder type DAC's, stray capacitance does not affect the speed of the response of the circuit due to constant ladder node voltages. Thus, it improves the speed performance.

- The major advantage of current mode D/A converter is that the voltage change across each switch is minimum. Thus, the charge injection is eliminated and the design of switch driver is simple.

Voltage Mode DAC

This is an alternative mode of DAC and it is called so because, its 2R resistance in the shunt path is switched between two voltages which is named as V_L and V_H. The output of this DAC is derived from the leftmost ladder node.

The input is sequenced through all the possible binary state starting from all 0's (0.....0) to all 1's (1.....1). The voltage of this node changes in steps of $2^{-n} (V_H - V_L)$ from minimum voltage of $V_o = V_L$ to maximum of $V_o = V_H - 2^{-n} (V_H - V_L)$.

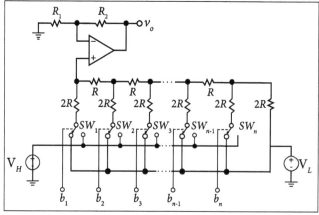

Voltage mode DAC.

The above diagram shows a non-inverting amplifier from which the final output is taken. Due to this buffering with non-inverting amplifier, scaling factor is given as,

$$K = 1 + (R_2 / R_1)$$

Advantages:

- More accurate selection and design of resistors R and 2R are possible and construction is simple.

- It allows us to interpolate between any two voltages which need not be a zero.

- The binary word length is increased by adding the required number or R-2R sections.

R - 2R Ladder

Wide range of resistors is needed in the binary weighted resistor type DAC. This may be avoided by using R-2R ladder type DAC. Here only two values of resistor are needed. It is best suited for integrated circuit realization.

The typical value of R range from 2.5kΩ to 10 kΩ:

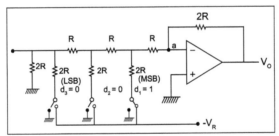

R-2R ladder DAC.

The above figure shows a 3-bit DAC where the switch positions d_1, d_2, d_3 corresponds to binary word 100.

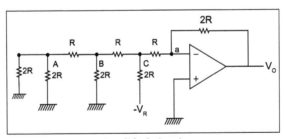

Simplified circuit.

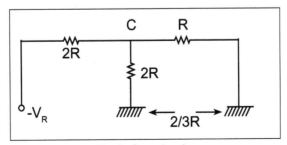

Equivalent circuit.

Voltage at node C is calculated by set procedure of the network analysis.

$$\frac{-V_R\left(\frac{2}{3}R\right)}{2R+\frac{2}{3}R}=\frac{-V_R}{4}$$

Output voltage is given as,

$$V_0=\frac{-2R}{R}\left(\frac{-V_R}{4}\right)=\frac{V_R}{2}=\frac{V_{FS}}{2}$$

The switch position that corresponds to the binary word 001 in 3-bit DAC is shown in the below figure:

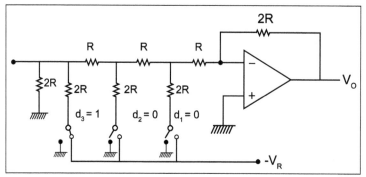

R-2R ladder DAC for switch position 001.

The voltage at the nodes (A, B, C) is formed by resistor branches which can be easily calculated in a similar fashion .Thus, the output voltage becomes,

$$V_0=\left(\frac{-2R}{R}\right)\left(\frac{-V_R}{16}\right)=\frac{V_R}{8}=\frac{V_{FS}}{8}$$

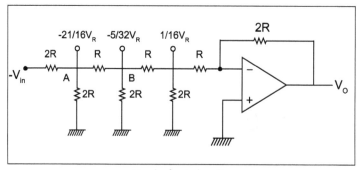

Equivalent circuit.

In a similar fashion, the output voltage for a R-2R ladder type DAC corresponding to other 3-bit binary words can be calculated.

Inverted R-2R

Inverted or Current Mode DAC

Current mode DAC's operates based on ladder currents. The ladder is formed by the resistance R in series path and the resistance 2R in shunt path. Thus the current is divided into i_1, i_2, i_3 ... in in each arm. The currents are either diverted to ground bus or to virtual-ground bus.

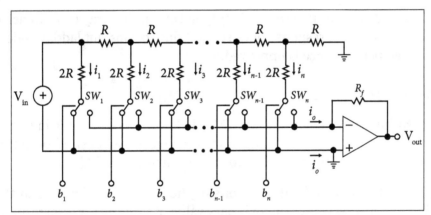

Inverted R-2R.

The currents are given as follows,

$$i_1 = V_{REF} / 2R = (V_{REF} / R)_{2-1}$$

$$i_2 = (V_{REF}) / 2 / 2R = (V_{REF} / R)_{2-2}$$

$$i_n = (V_{REF} / R)_{2-n}$$

Relationship between the currents can be given as,

$$i_2 = i_1 / 2$$

$$i_3 = i_1 / 4$$

$$i_4 = i_1 / 8$$

$$i_n = i_1 / 2n - 1$$

$$V_0 = -R_f i_0$$

$$V_0 = - (R_f / R) V_{REF} (b_{1(2-1)} + b_{2(2-2)} + ... + b_{n(2-n)})$$

The two currents i_o and $\overrightarrow{i_o}$ are complementary to each other and the potential of i_o

bus is close to the i_o bus. Linearity errors will occur. The final op-amp is used as current to voltage converter.

Advantages:

- The major advantage of current mode D/A converter is that voltage change across each switch is minimum. The charge injection is virtually eliminated and the switch driver design is simple.

- In current mode or inverted ladder type DAC's, the stray capacitance does not affect the speed of response of the circuit due to constant ladder node voltages. So speed performance is improved.

R-2R Ladder DAC

The enhanced binary-weighted resistor DAC is known as the R-2R ladder network. This type of DAC uses Thevenin's theorem to arrive at the desired output voltages. R-2R network has resistors with only two values namely R and 2R.

When each input is supplied either 0 volts or the reference voltage, then the output voltage will be an analog equivalent of binary value of three bits. V_{S2} corresponds to the most significant bit (MSB) and V_{S0} corresponds to the least significant bit (LSB).

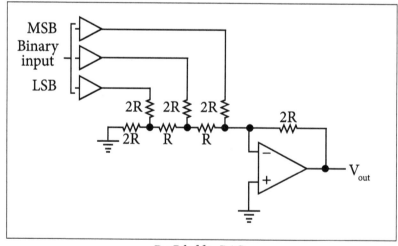

R-2R ladder DAC.

$$V_{out} = -\left(V_{MSB} + V_n + V_{LSB}\right) = -\left(V_{Ref} + V_{Ref}/2 + V_{Ref}/4\right)$$

R/2R DAC

An alternative to binary-weighted-input DAC is the R/2R DAC which uses fewer unique resistor values. A main disadvantage of the former DAC design was its requirement of various different precise input resistor values with one unique value per binary input bit.

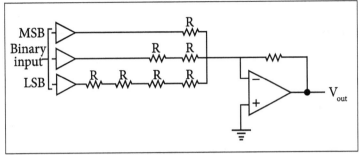

R/2R DAC.

This approach merely substitutes one type of complexity for another volume of components over diversity of component values. However, there is more efficient design methodology. By constructing a different kind of resistor network on the input of our summing circuit, we can achieve the same kind of binary weighting with only two kinds of resistor values and with only a modest increase in the resistor count. This ladder network is shown in the below figure:

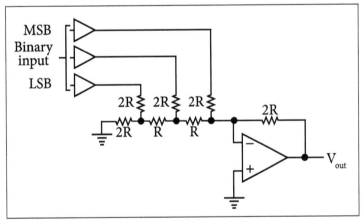

R/2R ladder DAC.

Ladder network is a bit more complex than for the previous circuit. Each input resistor provides an easily-calculated gain for that bit. Thevenin's theorem for each binary input and simulation program such as SPICE is used to calculate circuit response.

Binary	Output voltage
000	0.00 V
001	-1.25 V
010	-2.50 V
011	-3.75 V
100	-5.00 V

101	-6.25 V
110	-7.50 V
111	-8.75 V

If we are using +5 volts for a high voltage level and 0 volts for a low voltage level, then we can get an analog output directly corresponding to the binary input by using feedback resistance with a value of 1.6R instead of 2R.

4.7.1 ADC

Analog to Digital Converter (ADC) is an electronic integrated circuit used to convert the analog signals such as voltages to digital or binary form consisting of 1s and 0s.Most of the ADCs take a voltage input as 0 to 10V, -5V to +5V, etc. and correspondingly produces digital output as some sort of a binary number.

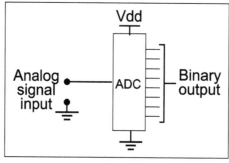

ADC Converter.

Analog to Digital Conversion Process

Analog to Digital Converter samples the analog signal on each falling or rising edge of sample clock. In each cycle, the ADC gets of the analog signal, measures and converts it into a digital value. The ADC converts the output data into a series of digital values by approximates the signal with fixed precision.

In ADCs, two factors determine the accuracy of the digital value that captures the original analog signal.

The main two steps involved in the process of conversion are:

 • Sampling and Holding.

 • Quantizing and Encoding.

The below figure shows the conversion of analog to digital. Bit rate decides decides the resolution of digitized output and we can observe in below figure where 3-bit ADC is used for converting analog signal.

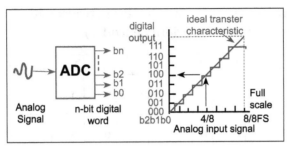

Analog to Digital Conversion Process.

Assume that one volt signal has to be converted from digital by using 3-bit ADC as shown below. Therefore, a total of $2^3 = 8$ divisions are available for producing 1V output. This results $1/8 = 0.125V$ is called as minimum change or quantization level represented for each division as 000 for 0V, 001 for 0.125, and likewise up to 111 for 1V. If we increase the bit rates like 6, 8, 12, 14, 16, etc. we will get a better precision of the signal. Thus, bit rate or quantization gives the smallest output change in the analog signal value that result from a change in the digital representation.

ADC is classified into five types as follows:

- Flash ADC.

- Counter type ADC.

- Successive approximation ADC.

- Dual slope ADC.

- Conversion times of typical ADC.

Flash or parallel comparator type:

This is the simplest possible A/D converter. It is the fastest and the most expensive technique. The below figure shows a 3-bit A/D converter.

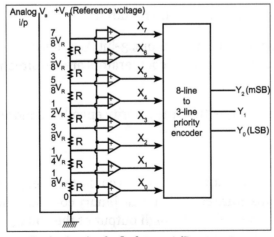

Basic circuit of a flash type A/D converter.

The above circuit consists of a resistive divider network with 8 op-amp comparators and a 8 line to 3 line encoder (3-bit priority encoder).

Input voltage (V_a)	X_7	X_6	X_5	X_4	X_3	X_2	X_1	X_0	Y_2	Y_1	Y_0
0 to $\dfrac{V_R}{8}$	0	0	0	0	0	0	0	1	0	0	0
$\dfrac{V_R}{8}$ to $\dfrac{V_R}{4}$	0	0	0	0	0	0	1	1	0	0	1
$\dfrac{V_R}{4}$ to $\dfrac{3V_R}{8}$	0	0	0	0	0	1	1	1	0	1	0
$\dfrac{3V_R}{8}$ to $\dfrac{V_R}{2}$	0	0	0	0	1	0	0	0	0	1	1
$\dfrac{V_R}{2}$ to $\dfrac{5V_R}{8}$	0	0	0	1	0	0	0	0	1	0	0
$\dfrac{5V_R}{8}$ to $\dfrac{3V_R}{4}$	0	0	1	1	1	1	1	1	1	0	1
$\dfrac{3V_R}{4}$ to $\dfrac{7V_R}{8}$	0	1	1	1	1	1	1	1	1	1	0
$\dfrac{7V_R}{8}$ to V_R	1	1	1	1	1	1	1	1	1	1	1

At each node of the resistive divider, voltage available at the nodes are equally divided between the reference voltage V_R and the ground. The purpose of the circuit is to compare the analog input voltage V_a with each of the node voltage. The truth table for the flash type A/D converter is shown in the above table.

Typical conversion time is 100ns or less which is limited by the speed of the comparator and the priority encoder. By using an advanced micro devices like AMD686A comparator and a T1147 priority encoder, conversion delays of the order of 20 ns is obtained.

Disadvantages:

- More number of comparators are required.

- A 2-bit ADC requires 3 comparators. 3-bit ADC needs 7, whereas, 4-bit require 15 comparator. 2n-1 number of comparators are required where n is the derived number of bits.

- If the value of n is large, then the priority encoder is more complex.

Counter Type ADC

In the below figure, the counter is reset to zero count by reset pulse. After releasing the reset pulse, the clock pulses are counted by the binary counter. These pulses go through the AND gate which is enabled by the high output of the voltage comparator. Number of counted pulses increase with time.

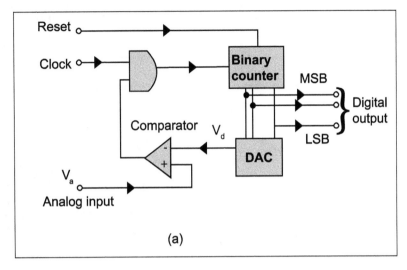

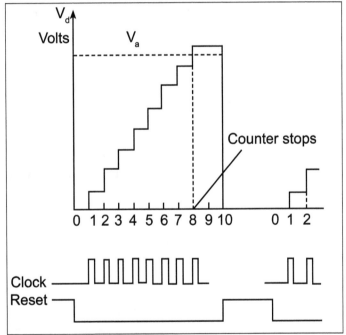

Counter type A/D converter.

The binary word representing this count is used as the input of a D/A converter whose output is a stair case. The analog output (V_d) of DAC is compared to the analog input input (V_a) by the comparator. If $V_a > V_d$, the output of the comparator is high and the AND gate allows the transmission of the clock pulses to the counter.

When $V_a < V_d$, the output of the comparator is low and the AND gate is disabled.

Dual Slope ADC

Dual slope conversion is an indirect method for A/D conversion where an analog voltage and reference voltage are converted into time periods by an integrator and then measured by counter.

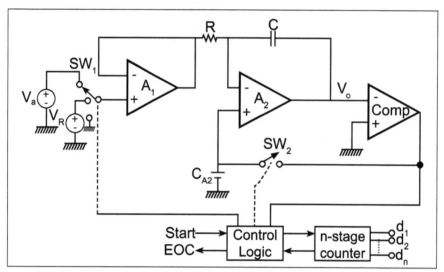

Functional diagram of the dual slope ADC.

The above figure shows the functional diagram of the dual slope or dual ramp converter. The analog part of the circuit consists of a high input impedance buffer A_1, precision integrator A_2 and a voltage comparator.

The converter integrates the analog input signal for a fixed duration of 2^n clock periods as shown in the below figure. It integrates an internal reference voltage (V_R) of opposite polarity until the integrator output is zero.

The number N of clock cycles required to reduce the integrator to zero is proportional to the value of V_a which is averaged over the integration period. Hence, N represents the desired output code.

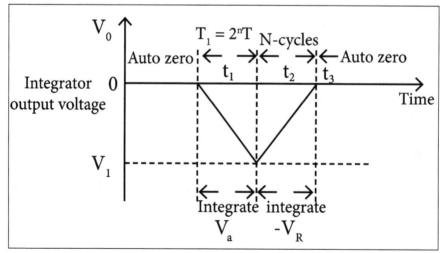

Integrated output waveform for the dual slope ADC.

Operation

Before the start command arrives, the switch SW_1 is connected to ground and SW_2 is closed. Any offset voltage present in the A_1, A_2, and comparator loop after integration appears across the capacitor till the threshold of the comparator is achieved.

The capacitor C_{AZ} provides automatic compensation for the input offset voltages of all the three amplifier. When SW_2 opens, C_{AZ} acts as a memory to hold the voltage which is required to keep the offset null. At the arrival of the START command at $t = t_1$, the control logic opens SW_2 and connects SW_1 to V_a and enables the counter starting from zero. This circuit uses an n-stage ripple counter and the counter is reset to zero after counting 2n pulses.

The analog voltage (V_a) is integrated for a fixed number 2^n counts of clock pulses after which the counter resets to zero. If the clock period is T, the integration takes place for a time $T_1 = 2^n \times T$ and the output is a ramp as shown in the above figure.

The counter resets itself to zero at the end of T1 and the switch SW1 is connected to the reference voltage $(-V_R)$. The output V_0 have a positive slope. As long as V_0 is negative, the output of the comparator is positive and the control logic counts the clock pulse.

When V_0 is zero at time $t = t_3$, the control logic has an end of conversion (EOC) command and clock pulses do not enter the counter. Reading of the counter at t_3 is proportional to the analog output voltage V_a.

$$T_1 = t_2 - t_1 = \frac{2^n \text{ counts}}{\text{Clock rate}}$$

$$t_3 - t_2 = \frac{\text{digital count N}}{\text{Clock rate}}$$

For an integrator, we have,

$$\Delta V_0 = \left(-\frac{1}{RC}\right) V(\Delta t)$$

Voltage (V_0) is equal to V_1 at the instant t_2 which is written as,

$$V_1 = \left(-\frac{1}{RC}\right) V_a (t_2 - t_1)$$

Voltage V_1 is given as,

$$V_1 = \left(-\frac{1}{R_C}\right) - V_R (t_2 - t_3)$$

$$V_a\left(t_2-t_1\right)=V_R\left(t_3-t_2\right)$$

$$\left(t_2-t_1\right)=2^n$$

$$\left(t_3-t_2\right)=N$$

$$V_a\left(2^n\right)=V_R\left(N\right)$$

$$V_a=\left(V_R\right)\left(N/2^n\right)$$

Since V_R and n are constant, the analog voltage V_a is proportional to the count reading N and it is independent of R, C and T.

The dual slope ADC integrates the input signal for a fixed time. Hence, it provides excellent noise rejection of AC signals whose periods are integral multiples of the integration time T_1.

The main disadvantage of the dual slope ADC is the long conversion time. Dual slope converters are suitable for accurate measurement of slowly varying signals such as thermocouples and weighing scales. Dual slope ADC is the basis of digital panel meter and multi meter.

Successive Approximation Type ADC

The successive approximation technique uses a very efficient code search strategy to complete n-bit conversion in just n clock periods.

An eight bit converter require 8 clock pulses to obtain a digital output. The below figure shows an 8-bit converter which uses a successive approximation register(SAR) to find the required EOC value of each bit by trial and error method.

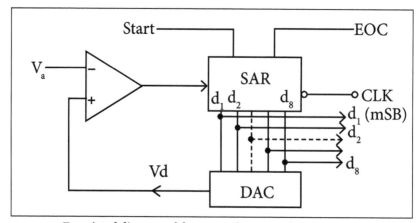

Functional diagram of the successive approximation ADC.

Operation

With the arrival of the start command, the SAR sets the MSB d_1 = 1 where all other bits are zero and the trial code is 10000000. The output V_d of DAC is compared with the analog output V_a. If V_a is greater than the DAC output V_d, then 10000000 is less than the correct digital representation.

MSB at left is '1' and the next lower significant bit is 1. If V_a is less than the output of DAC, then 10000000 is greater than the correct digital representation. MSB is reset to 0 and next significant bit is considered. This procedure is repeated for all subsequent bits one at a time until all bit position have been tested.

When the DAC output crosses V_a, the comparator changes state and this can be taken as the end of conversion (EOC) command. The below table shows a typical conversion sequence.

Successive approximation conversion sequence for a typical input is shown in the below table:

Correct digital representation	Successive approximation register output V_d at different stages in the conversion	Comparator output
11010100	10000000	1 [Initial output]
	11000000	1
	11100000	0
	11010000	1
	11011000	0
	11010100	1
	11010110	0
	11010101	0

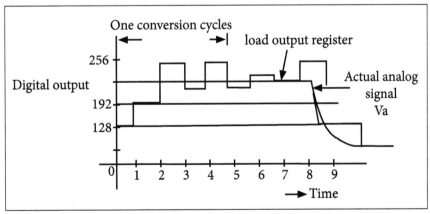

Output characteristics of D/A.

D/A output voltage is successively closer to the actual analog input voltage. It requires 8 pulses to establish the accurate output regardless of the value of the additional analog input. However, one circuit pulse is used to load the output register and reinitialize the circuit.

Measurements and Instrumentation

5.1 Transducers and Classification of Transducers

Transducer

An electrical transducer is a device which converts the physical quantity into a proportional electrical quantity such as voltage or current. Hence it converts any quantity to be measured into usable electrical signal. Physical quantity may be pressure, level, temperature, displacement etc.,

The output obtained from the transducer is in the electrical form and it is equivalent to the measured quantity. For example, a temperature transducer will convert temperature to an equivalent electrical potential. This output signal is used to control the physical quantity or to display it.

Any device which convert one form of energy into another form is called as a transducer. For example, a speaker is known as a transducer as it converts electrical signal to pressure waves or sound. But an electrical transducer will convert a physical quantity to an electrical one.

A device which converts a physical quantity into the proportional electrical signal is called as an electrical transducer.

Classification of Electrical Transducer

1. Transducer broadly classified into two groups based on its internal property:

- Active transducer: These transducers are self-generating type transducers which develop their own voltage or current.

Example: Thermocouples, photovoltaic cell, etc.,

- Passive transducer: They are externally powered transducer which absorb some energy from physical phenomenon under steady.

Example: Photoemission cell, thermistor, etc.

2. Classification based on type of output:

- Analog transducer: There transducers convert the input physical phenomenon into analogous output which is a continuous function of time.

Example: Strain gauge, LVDT, etc.,

- Digital transducer: These transducers convert the input physical phenomenon into digital output which may be in the form of pulse.

3. Classification based on electrical principle involved:

- Variable-resistance type:
  - Stain gauges and pressure gauges.
  - Thermistors, resistance thermometers.
  - Photovoltaic cells.
- Variable-inductance type:
  - Linear voltage differential transformer (LVDT).
  - Reluctance pick-up.
  - Eddy current gauge.
- Variable-capacitance type:
  - Capacitor microphone.
  - Pressure gauge.
  - Dielectric gauge.
- Voltage-generating type:
  - Thermocouple.
  - Photovoltaic cell.
  - Rotational motion tachometer.
  - Piezoelectric pickup.
- Voltaic-divider type:
  - Potentiometer position sensor.
  - Pressure - actuated voltage divider.

5.1.1 Resistive Transducer

The three passive elements in an electric circuits are resistor, inductor and capacitor.

The transducers that are based on the variation of the parameters due to application of any external stimulus are known as passive transducers.

Resistive transducer is a transducer in which a variation in a quantity or signal produces a variation in resistance which in turn produces a proportional conversion to a quantity or signal in another form.

For example, in a resistance thermometer, a change in temperature causes a change in the resistance of the resistive element which in turn produces a signal that is interpreted for readings. It is also called as resistance transducer.

Working Principle

Transducer contains two parts that are closely related to each other i.e., the sensing element and transduction element.

The sensing element is called as sensor. It is a device which produces measurable response to change in physical conditions. The transduction elements convert the sensor output to suitable electrical form.

The dc resistance R can be given by,

$$R = \frac{\rho l}{a}$$

where,

l = Conductor length in m

a = Area of cross section in m²

ρ = Specific resistivity in Ω-m

Change in the value of the resistive element can be brought about by subjecting the element to external stimulus that affects either dimensions of the clement or its resistivity. The dimensional changes can be brought about by subjecting the resistive elements, to pressure, force or torque, directly or by means of some primary transducers. Resistive strain gauges enable measurement of strain of mechanical members on to which they are bonded. The resistivity of the material medium constituting the path of the current in the resistor varies with the temperature and composition of the medium. Resistance thermometers are known to be exceptionally good for temperature measurements.

5.1.2 Inductive Transducer

Inductance is a measure that relates electrical flux to current. Inductance reactance is a measure of the inductive effect and it is expressed as,

$$X = 2pfL$$

Where,

f is the frequency of the applied voltage in Hz.

X is the inductive reactance in ohm's.

L is the inductance in Henry.

The inductance of a circuit is influenced by the following factors:

- Size of the coil.

- Permeability of the flux path.

- Number of turns in a coil.

As a result of a mechanical displacement, the permeability of the flux path is altered and the inductance of the system changes. Inductance is monitored through the resonant frequency of the inductance coils to an applied voltage.

As inductance changes, the resonant frequency of the coils changes. Electronic circuits that convert frequency to voltage are used to gain a voltage output to inductive transducers.

Inductive transducers can be classified as follows:

- Variable self-inductance (Single coil, two coil).

- Variable mutual inductance (Simple two coil, three coil).

- Variable reluctance (Moving iron, moving coil, moving magnet).

The various arrangements that lead to the above classifications are shown in the below figures a, b and c:

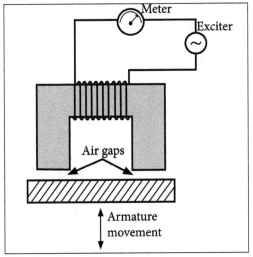

(a) Simple self-inductance arrangement where the change in the air gap changes the pickup output.

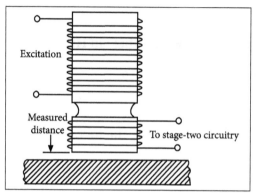

(b) Two coil inductive pickup for an electronic micrometer.

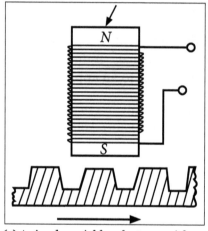

(c) A simple variable reluctance pickup.

Variable-reluctance Transducers

There are two major parts such as a ferromagnetic rotor and a stator assembly. In the stator, four coils a, b, c and d are connected together with the voltages induced in coils a and c which is same as the voltages induced in coils b and d at NULL position of the ferromagnetic rotor.

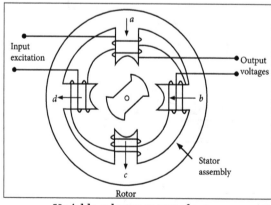

Variable-reluctance transducers.

Based on the rotation of the rotor in clockwise direction, the reluctance will increase in the coils a and c which decreases the reluctance in the coils b and d which gives a net output voltage (k). If the rotation is in counter clockwise direction, it produces same kind of effect as in coils b and d with 180° phase shift.

It is possible to detect very small motion which provides output signal for even 0.01 degree of changes in angles. They have the sensitivity as high as five volt per degree rotation.

5.1.3 Capacitive Transducer

The capacitive transducer is nothing but the capacitor with variable capacitance.

Principle

The capacitance of the capacitor varies as the material comes between two plates of capacitor and the presence of something is measured by capacitance.

Capacitive Pressure Transducers

When the distance between the two parallel plates changes, capacitance of the parallel plate capacitor changes. The capacitive pressure transducer is shown in the below figure:

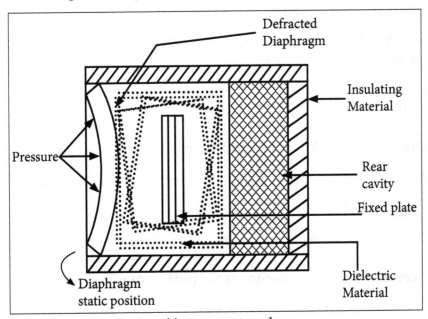

Capacitive pressure transducer.

In this transducer, diaphragm acts as one of the plates of a two plate capacitor while other plate is fixed. The fixed plate and the diaphragm are separated by a dielectric material. When the force is applied to the diaphragm, it changes its position from initial static position where no force is applied. Due to this, the distance of separation between the fixed plate and the diaphragm changes and the capacitance also changes.

The change in the capacitance is measured by using any simple AC bridge. But practically, the change in capacitance is measured using an oscillator circuit where capacitive transducer is part of that circuit. When capacitance changes, the oscillator frequency changes. By using capacitive transducer, applied force is measured in terms of change in the capacitance.

Advantages:

- Power required to operate is small.

- It is highly sensitive.

- It has high input impedance and good frequency response.

- It is useful in the applications where stray magnet fields affect the performance of an inductive transducer.

Disadvantages:

- Proper insulation is required.

- Stray capacitances affect the performance of the transducer.

- It shows non-linear behavior.

- Due to long leads and the cables used, loading effect makes frequency response poor and reduces sensitivity.

- For low value capacitance, the output impedance has high values which causes loading effects.

Capacitance of Capacitive Transducers

The capacitance C between the two plates of capacitive transducers is given as,

$$C = \varepsilon_{ox} \, \varepsilon_{rx} \, A / d$$

Where,

C - Capacitance of the capacitor or the variable capacitance transducer.

ε_o – Absolute permittivity.

ε_r – Relative permittivity.

The product of ε_o and ε_r is also known as the dielectric constant of the capacitive transducer.

D - Distance between the plates

A - Area of the plates

Capacitance of the capacitive transducer depends on the area of the plates and also on the distance between the plates. Capacitance of the capacitive transducer changes with the dielectric constant of the dielectric material used in it.

Capacitance of the variable capacitance transducer changes with the change in the dielectric material, change in the area of the plates and the distance between the plates.

5.1.4 Thermoelectric Transducer

A thermocouple is a temperature-measuring device which consist of two dissimilar conductors that contact each other at one or more spots.

A thermocouple comprises of at least two metals which are joined together to form two junctions.

The other junction is connected to a body of known temperature. This is the cold or reference junction. Therefore, the thermocouple measures unknown temperature of the body with reference to the known temperature of the other body.

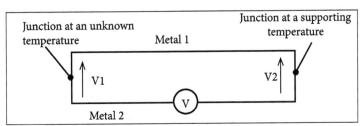

Principle of thermocouple operation.

Working Principle

The working principle of thermocouple depends on three effects namely See beck, Peltier and Thomson effect.

They are as follows:

See Beck Effect

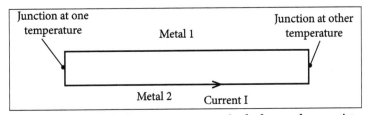

See beck effect at which the current emerges in the layout that consists of junctions from two heterogeneous metals at different temperature.

It states that when two different or unlike metals are joined together at two junctions, an electromotive force (emf) is generated at the two junctions. The amount of emf generated is different for different combinations of the metals.

Peltier Effect

As per the Peltier effect, when two dissimilar metals are joined together to form two junctions, emf is generated within the circuit due to the different temperatures of the two junctions of the circuit.

Thomson Effect

As per the Thomson effect, when two unlike metals are joined together to form two junctions, the potential exists within the circuit due to temperature gradient along the entire length of the conductors within the circuit.

The emf suggested by the Thomson effect is very small and it can be neglected by making proper selection of the metals. Peltier effect plays a prominent role in the working principle of the thermocouple.

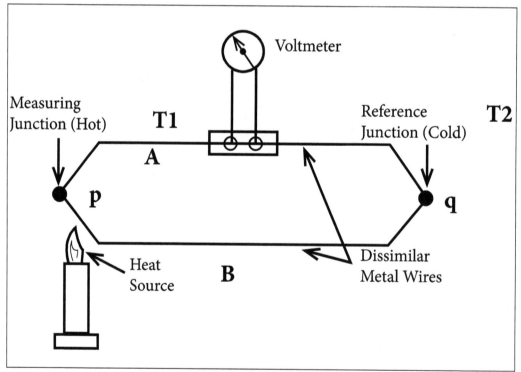

Thermocouple circuit.

Working of Thermocouple Circuit

The general circuit for the working of thermocouple is shown in the above figure. It comprises of two dissimilar metals namely A and B. They are joined together to form two junctions namely p and q which are maintained at the temperatures T_1 and T_2 respectively.

Thermocouple is not formed if there are no two junctions. Since the two junctions are maintained at different temperatures, the Peltier emf is generated within the circuit and it is the function of the temperatures of two junctions.

If the temperature of both the junctions are same, equal and opposite emf will be generated at both junctions and the net current flowing through the junction is zero. If the junctions are maintained at different temperatures, the emf's will not be zero and net current flows through the circuit.

The total emf flowing through this circuit depends on the metals used in the circuit as well as the temperature of the two junctions. The total emf or the current flowing through the circuit is be measured by the suitable device.

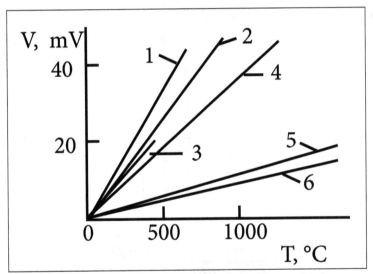

Typical dependencies of voltage on difference of temperature for thermocouple.

In the thermocouple circuit, junctions are made up of different metals as follows:

- Nickel-chromium/constantan (E type).

- Iron/constantan (J type).

- Copper/constantan (T type).

- Nickel-chromium/nickel-manganese-aluminum-silicon (K type).

- Platinum-rhodium/platinum (R type).

- Platinum/rhodium/platinum (S type).

Applications of thermocouple circuit:

- Thermocouples are used as temperature sensors for measurement, control and it converts temperature into electricity.

- They are used in research and industry and temperature measurement for furnaces, gas turbine exhaust, diesel engines and other industrial processes.

- They are also used in homes & offices as temperature sensors in thermostats and also in flame sensors for fire detection.

5.1.5 Piezoelectric Transducer

In certain materials called piezoelectric materials a potential difference appears across their opposite faces as a result of dimensional changes due to application of pressure created by mechanical force. This potential is produced as a result of displacement of charges in the body of the material. The effect is reversible, i.e., the reverse happens when a varying potential is applied to the proper axis of the crystal; a change in the dimensions of the crystal occurs. This effect is known as the piezoelectric effect.

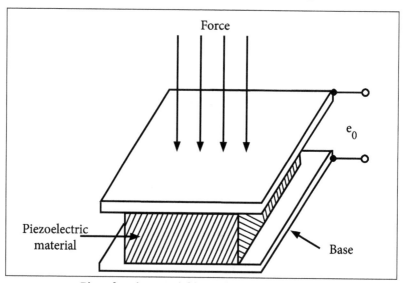

Piezoelectric material is used to measure force.

Commonly used piezoelectric materials are quartz, barium, titanite, lithium sulphate, etc. The above figure shows a piezoelectric material used for the measurement of force. Since these transducers are self-generating, i.e., active transducers, the EMF induced due to force applied is directly proportional to the force. As shown in the above figure, an external force, which is to be measured, exerts a pressure on the top of the crystal and as a result EMF is produced across the crystal.

A piezoelectric material should not be sensitive to temperature and humidity variations. They should lend themselves to forming different shapes. Quatz is the most suitable piezoelectric material on this account. However, the voltage induced is quite small. Rochelle salt provides higher values of induced EMF, but it is affected by temperature variations.

A piezoelectric material should not be sensitive to temperature and humidity variations. They should lend themselves to forming different shapes. Quatz is the most suitable piezoelectric material on this account. However, the voltage induced is quite small. Rochelle salt provides higher values of induced EMF, but it is affected by temperature variations. The below figure shows a pressure transducer which utilizes the property of piezoelectric crystals. The transducer consists of a diaphragm by which pressure is transmitted to the piezoelectric crystal. The crystal generates an EMF across its two surfaces which is proportional to the magnitude of the applied pressure.

Piezoelectric pressure transducers are used to measure high pressure that changes rapidly like the pressure inside a cylinder of a petrol or diesel engine, or a compressor. The main drawback of this transducer is that the output voltage in affected by temperature variations of the crystal.

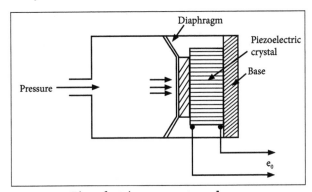

Piezoelectric pressure transducer.

Advantages:

- Its construction is rugged and size is small.

- Piezo electric output is high with negligible phase shift.

- Its frequency response is excellent.

Application of piezoelectric transducer:

- It is used to measure non-electric quantities such as acceleration, vibration, pressure etc.

- It is widely used in aerodynamics, supersonic wind tunnels etc.

- It is used in ultrasonic, non-destruction test equipment.

- It is also used in spark ignition engines.

5.1.6 Photoelectric Transducer

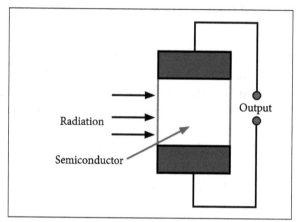

Photoelectric transducer.

The photoelectric transducer converts the light energy into electrical energy. It is made up of semiconductor material. The photoelectric transducer uses a photosensitive element which ejects the electrons when the beam of light is passes through it.

The below figure shows a light source and a photo tube used for the measurement of pressure. The output voltage depends upon the amount of light falling on the tube through the window. The opening of the window is con-trolled by the pressure of the gas falling on a membrane. The output voltage approximately varies linearly with the displacement of the aperture, and hence the pressure.

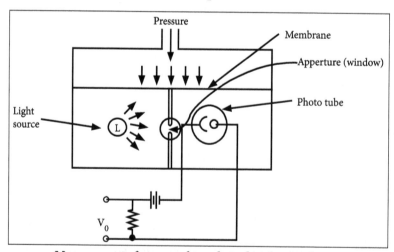

Measurement of pressure by a photoelectric transducer.

Classification of Photoelectric Transducer

A photoelectric transducer can be categorized as photo emissive, photoconductive or photovoltaic. In photo emissive devices, radiation falling on a cathode causes electrons to be emitted from the cathode surface.

In photoconductive devices, the resistance of a material is changed when it is illuminated. Photovoltaic cells generate an output voltage which is proportional to the radiation intensity. The incident radiation may be infrared, ultraviolet, gamma rays or X-rays as well as visible light.

Photo Emissive Cell

The Photo emissive cell converts the photons into electric energy. It consists of an anode rod and cathode plate. The anode and cathode are coated with a photo emissive material called as Caesium antimony.

When the radiation of light fall on cathode plates the electrons starts flowing from anode to cathode. Both the anode and the cathode are sealed in a closed, opaque evacuated tube. When the radiation of light fall on the sealed tube, the electrons starts emitting from the cathode and moves towards the anode. The magnitude of the current is directly proportional to the intensity of light passed through it.

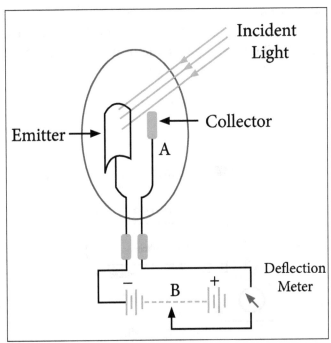

Photo emissive cell.

Photoconductive Cell

It converts the light energy into an electric current. It uses the semiconductor material like cadmium selenide, Ge, Se as a photo sensing element.

When the beam of light falls on the semiconductor material, their conductivity increases and the material works like a closed switch. The current starts flowing into the material and deflects the pointer of the meter.

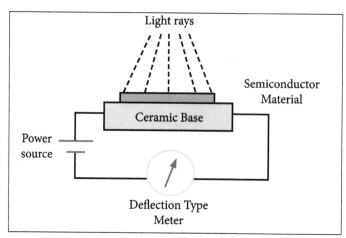

Photoconductive cell.

Photovoltaic Cell

The photovoltaic cell is a type of active transducer. The current starts flowing into the photovoltaic cell when the load is connected to it. The silicon and selenium are used as a semiconductor material. When the semiconductor material absorbs heat, the free electrons of the material starts moving. This phenomenon is known as the photovoltaic effect.

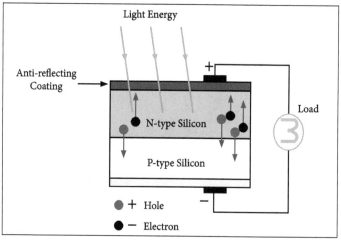

Photovoltaic cell.

The movements of electrons develop the current in the cell and the current is known as the photoelectric current.

5.1.7 Hall-effect Transducer

The principle of working of a Hall effect transducer is that when a strip of conducting material carries current in the presence of the transverse magnetic field, as shown in the below figure, an EMI' will he induced between the opposite edges of the conducting strip. The magnitude of the voltage induced too will depend upon the material of the strip, the current, and the magnetic field strength. Thus, we can state that Hall effect refers to the potential difference (Hall voltage) on the opposite sides of an electrical conductor through which an electric current is flowing, created by a magnetic field perpendicular to the current.

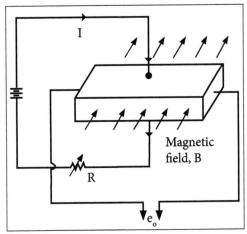

Hall-effect transducer.

Hall coefficient is defined as the ratio of the output voltage to the product of the current and magnetic field, i.e., $e_o / (I \times B)$ divided by the thickness of the element, t.

Thus,

$$\text{Hall coefficient} = \frac{e_o}{I \times B \times t}$$

Hall coefficient is the characteristic of the material, i.e., it depends on the material from which the conductor is made. To understand how a potential gets developed due to Hall effect, let us consider the following explanation.

When current flows through the conductor, it consists of the movement of charge carriers. In the presence of a perpendicular magnetic field, the moving charges experience a force, called the Lorenz force. This makes the path of the moving charge somewhat curved (and not a straight line flow) so that charges accumulate on one face of the conductor material. Equal and opposite charges start appearing on the other face of the conductor material where there is a shortage of mobile charges. This results in unequal and asymmetric distribution of charges across the conducting element both perpendicular to the direction of a straight line flow of charge (in the absence of a magnetic field) and in the direction of the applied magnetic field. This separation of charge establish an electric field that would oppose the migration of further charge, and therefore an electrical potential gets build up.

Hall effect transducers are non-contact devices, have small size, and high resolution. Such transducers can be used in the measurement of velocity, revolutions per second, magnetic field, charge carrier density, measurement of displacement, etc. Hall effect transducers can be used to measure current in a conductor without actually connecting a meter in the conducting circuit.

5.2 Mechanical Classification of Instruments: Types of Indicating Instruments and Multi Meters

Classification of Measuring Instruments

The instrument used for measuring the physical and electrical quantities is known as the measuring instrument.

The measuring instrument is classified into three types as follows:

- Electrical instrument.

- Electronic instrument.

- Mechanical instrument.

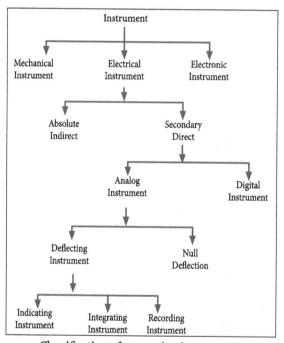

Classification of measuring instrument.

Mechanical Instrument

It is used for measuring the physical quantities. This instrument is suitable for measuring the static and stable condition because the instrument is unable to give the response to the dynamic condition. The electronic instrument has quick response time.

The instrument provides the quick response when compared to the electrical and mechanical instrument.

Electrical Measuring Instrument

The electrical instrument is used for measuring electrical quantities likes current, voltage, power, etc. The ammeter, voltmeter, wattmeter are the examples of the electrical measuring instrument. The ammeter measures the current in amps; voltmeter measures voltage and Wattmeter are used for measuring the power. The classification of the electric instruments depends on the methods of representing the output reading.

Electrical measuring instruments are mainly classified as follows:

- Indicating instruments.

- Recording instruments.

- Integrating instruments.

Indicating Instrument: It uses a dial and pointer to show or indicate the magnitude of an unknown quantity. The examples are ammeters, voltmeter etc.

Recording Instruments: These instruments give a continuous record of the given electrical quantity which is measured over a specific period.

In recording instruments, the readings are recorded by drawing the graph. The pointer of such instruments is provided with a marker i.e., pen or pencil which moves on graph paper as per the reading. X-Y plotter is the best example of such an instrument.

Integrating Instruments: These instruments measure the total quantity of electricity delivered over a period of time. For example, a household energy meter registers number of revolutions made by the disc to give the total energy delivered with the help of counting mechanism consisting of dials and pointers.

5.2.1 Types of Indicating Instruments

Galvanometer Type Recorder

D'Arsonval movement used in moving coil indicating instruments provide the movement in a galvanometer recorder. D'Arsonval movement consists of a moving coil which is placed in a strong magnetic field as shown in the below figure (a).

In a galvanometer type recorder, the pointer of the D'Arsonval movement is fitted with a pen-ink (stylus) mechanism. The pointer deflects when current flows through the moving coil. The deflection of the pointer is directly proportional to the magnitude of the current flowing through the coil. As the signal current flows through the coil, the magnetic field of the coil varies in intensity in accordance with the signal.

The reaction of this field with the field of the permanent magnet causes the coil to change its angular position. As the position of the coil follows the variation of the signal current being recorded, the pen is accordingly deflected across the paper chart. The paper is pulled from a supply roll by a motor driven transport mechanism. Thus, as the paper moves past the pen and as the pen is deflected, the signal waveform is traced on the paper.

The recording pen is connected to an ink reservoir through a narrow bore tube. Gravity and capillary action establishes a flow of ink from the reservoir through the tubing and into the hollow of the pen.

Galvanometer type recorders are well suited for low frequency ac inputs obtained from quantities varying slowly at frequencies of up to 100 c/s or in special cases up to 1000 c/s. Because of the compact nature of the galvanometer unit (or pen motor), this type of recorder is suitable for multiple channel operation.

Hence, it finds extensive use in the simultaneous recording of a large number of varying transducers outputs. This recorder uses a curvilinear system of tracing. The time lines on the chart must be arcs of radius R (where R is the length of the pointer) and the galvanometer shaft must be located exactly at the center of curvature of a time line arc.

Improper positioning of the galvanometer or misalignment of the chart paper in the recorder will give a distorted response. One method of avoiding the distorted appearance of recordings in curvilinear coordinates is to produce the recording in rectangular coordinates.

The chart paper is pulled over a sharp edge that defines the locus of the point of contact between the paper and the recording stylus. The stylus is rigidly attached to the galvanometer coil and wipes over the sharp edge as the coil rotates.

In one of the recorders, the paper is heat sensitive and the stylus is equipped with a heated tip regardless of the stylus position on the chart.

The paper can be electrically sensitive where the stylus tip-carry current into the paper at the point of contact. The recorders can work on ranges ranging from a few mA/mV to several mA/mV.

These moving galvanometer type recorders are inexpensive instruments which have a narrow bandwidth of 0-10 Hz. They have sensitivity of about 0.4 V/mm or from a chart of 100 mm width, a full scale deflection of 40 mV is obtained.

In most instruments, the speed of the paper through the recorder is determined by the gear ratio of the driving mechanism. If it is desired to change the speed of the paper, one or more gears must be changed.

Paper speed is an important consideration for several reasons.

They can be given as:

- If the paper moves too slowly, the recorded signal variations are bunched up and it is difficult to read.

- If the paper moves too fast, the recorded waveform will spread out at that greater lengths of paper which will record the variations of the signal. It also makes the task of reading and interpreting the waveforms more difficult.

- The operator determines the frequency components of the recorded waveform. The paper is usually printed with coordinates such as graph paper.

Some recorders contain a timing mechanism that prints a series of small dots along the edge of the paper chart where the paper moves through the recorder. This time marker produces one mark per second.

These types of recorders are mostly used as optical recorders and contain a light source provided by either an ultra violet or tungsten lamp.

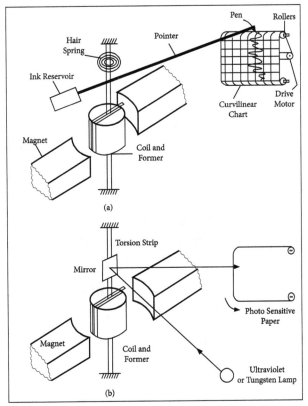

(a) Galvanometer type recorder (b) Optical galvanometer recorder.

A small mirror is connected to the galvanometer movement and the light beam is focussed on this mirror as shown in the below figure (b). The beam reflected from the mirror is focussed into a spot on a light sensitive paper.

As the current passes through the coil, the mirror deflects. The movement of the light beam is affected by the deflection of the small mirror and the spot on the paper also varies for the same reason to trace the waveform on the paper.

5.2.2 Multi Meters

DVM is an analog to digital converter (A/D) with a digital display voltmeter.

Digital multi meter (DMM) = Electronic volt ohm millimeter with digital display

DMM is one of the most common items of test equipment used in the electronics industry today. Though there are many other items of test equipment that are available, the multi meters provide excellent readings of the basic measurements of amps, volts and ohm's.

Digital multi meters use digital and logic technology rather than analogue techniques which enables many new test features to be embedded in the design.

Types of DVM's are given below as follows:

- Ramp type DVM.

- Integrating type DVM.

- Potentiometric type DVM.

- Successive approximation type DVM.

- Continuous balance type DVM.

Dual Slope Integrating Type Digital Voltmeter

In this type, the most popular method of analog to digital conversion is used. The basic principle is that the input signal is integrated for fixed interval and then same integrator is used to integrate the reference voltage with reverse slope.

When the switch S_1 is in position 1, the capacitor starts charging from zero. The rate of charging is proportional to the input voltage level.

The output of the op-amp is given as,

$$V_{out} = \frac{1}{R_1 C} \int^{V_1} V_{in} \, dv$$

$$V_{out} = \frac{V_{in}}{R_1} \frac{t_1}{C} \qquad ...(1)$$

After time interval, the input voltage is disconnected and a negative voltage $-V_{ref}$ is connected by throwing the switch in position 2.

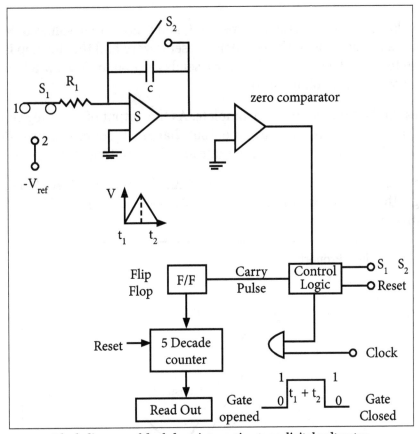

Block diagram of dual slope integrating type digital voltmeter.

In this position, the output of op-amp is given as,

$$V_{out} = \frac{1}{R_1 C} \int_0^{} -V_{ref} \, dv$$

$$V_{out} = \frac{-V_{ref} \, t_2}{R_1 C} \qquad ...(2)$$

Subtracting (1) from (2), we have,

$$V_{out} - V_{out} = 0 = \frac{V_{ref} \, t_2}{R_1 C} - \left(\frac{V_{in} \, t_1}{R_1 C} \right)$$

$$\frac{V_{ref}\, t_2}{R_1 C} = \frac{V_{in}\, t_1}{R_1 C}$$

$$V_{in} = V_{ref}\left(\frac{t_2}{t_1}\right)$$

The input voltage depends on the time period t_1 and t_2 not on the values of R_1 and C. At the start of the measurement, the counter is reset to zero and the flip-flop is also zero. This is given to control logic and this control sends a signal to close switch to position 1 and integration of input voltage starts.

It continues till the time period is incomplete. As the output of the integrator changes from its zero value, the zero comparator output changes its state. This provides a signal to control logic which in turns opens the gate and the counting of lock pulse starts.

If the counter exceeds 9999, then the output of the flip-flop is activated to logic level 1. This changes the switch position from S_1 to S_2. Then $-V_{ref}$ is connected to op-amp which gives negative slope.

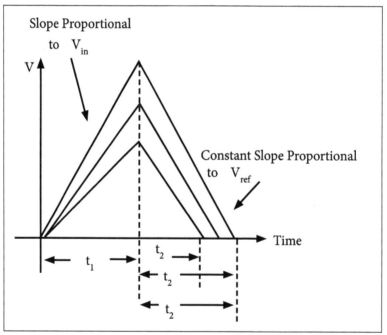

Basic principle of dual slope method.

The output decreases linearly and attains zero value when the capacitor C is fully discharged. Gate is closed after time t_1 or t_2,

$$V_{in} = V_{ref}\left(\frac{t_2}{t_1}\right)$$

Ramp Type Digital Voltmeter

Linear ramp voltage changes from input voltage to zero voltage or vice versa. This time interval is measured with an electronic time interval counter and the count is displayed as a number of digits on electronic indicating tubes of the output readout of the voltmeter.

The conversion of a voltage value of a time interval is shown in the below timing diagram.

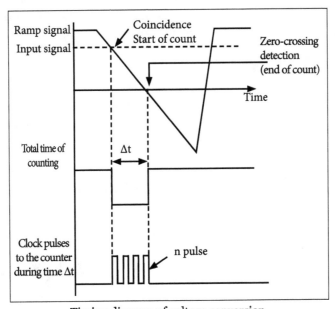

Timing diagram of voltage conversion.

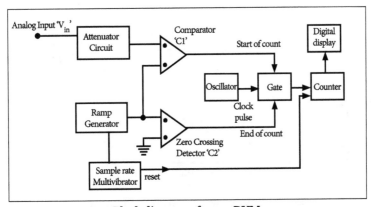

Block diagram of ramp DVM.

At the start of measurement, a ramp voltage is initiated. A negative ramp is shown in the above figure where a positive going ramp is used. The ramp voltage value is continuously compared with the measured voltage. At the instant, the value of ramp voltage is equal to that of unknown voltage at a coincidence circuit is called as an input comparator which generates a pulse that opens a gate.

The ramp voltage continues to decrease till it reaches ground level. At this instant, another comparator called as ground comparator generates a pulse and closes the gate. The time clasped between opening and closing of the gate is 't' as indicated in the above figure. During this time interval, pulses from a clock pulse generator pass through the gate and they are counted and displayed.

The decimal number as indicated by the readout is the measure of the value of input voltage. The sample rate multi vibrator determines the rate at which the measurement cycles are initiated. The sample rate circuit provides an initiating pulse for the ramp generator to start its next ramp voltage. At the same time, it sends a pulse to the counter which sets all of them to 0. This momentarily removes the digital display of the readout.

Digital Multi-meter

Digital meters offer higher accuracy, high input impedance, unambiguous readings at greater viewing distances, smaller size and a digital electrical output in addition to visual readout.

The main part of most of the digital multi-meter (DMM's) is the analog to digital converter (A/D) which converts an analog input signal to a digital output.

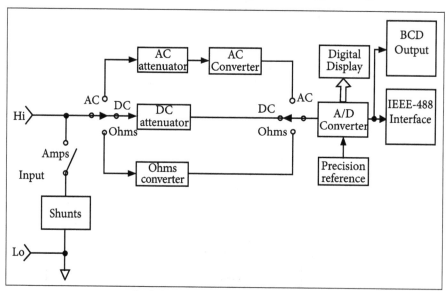

Block diagram of digital multi-meter.

Since the DMM is a voltage sensing meter, current is converted to volts by passing it through a precision low resistance shunt where AC is converted to DC at the AC converter by employing rectifiers and filters.

Most of the AC converters detect the peak value of the signals and they give the rms value of a sine wave. Finally, this DC level is applied to the A/D converter to obtain the digital information.

For resistance measurement, the meter includes a precision low current source that is applied across the unknown resistor. DC voltage drop across the resistor is proportional to the value of the unknown resistor.

For AC measurements, the digital multi-meter is a true rms instrument which measures true rms value of any periodic signal.

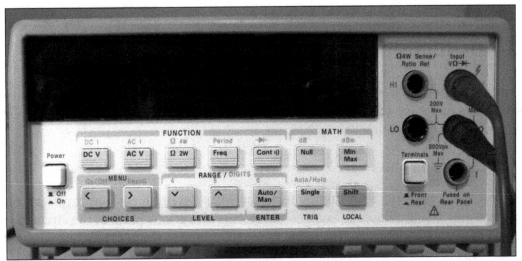

Digital multi-meter in the laboratory.

Multi-meter Operation

The operation of the analog multi-meter is quite easy. If the meter is new, then we should install any battery needed for the resistance measurements.

Steps:

- Insert the probes into the correct connections.

- Set the switch to the correct measurement type and range for the measurement is made. When selecting the range, ensure that the maximum range is above that anticipated. The range on the multi-meter is reduced later if necessary. By selecting the range i.e., too high, it prevents overloading of the meter and any possible damage to the movement of the meter itself.

- Maximum deflection of the meter is adjusted.

- Once reading is complete, it is a wise precaution to place the probes into the voltage measurement sockets and turn the range to maximum voltage. In this way, if the meter is accidentally connected without thinking about the range used, there is little chance of damage to the meter. This may not be true if it left set for a current reading and the meter is accidentally connected across a high voltage point.

Advantages:

- Accuracy is very high.

- Input impedance is very high.

- Size is compact and cost is low.

5.3   Oscilloscopes

An oscilloscope displays waveforms in a two dimensional format. The vertical axis is used to plot incoming voltage and the horizontal axis is used as a time axis. Waveform voltage is a function of time.

Oscilloscope Probes

Oscilloscope is used for fault finding for electronics development and repair or diagnostics work. The oscilloscope enables waveforms on various parts of the circuit to be viewed in a graphical format. To enable the oscilloscope to connect to required points, oscilloscope probes or scope probes are required.

A signal line is used and earth return connection to form a simple oscilloscope probe. This approach does not provide optimum performance since both the electrical and mechanical aspects meet the necessary requirements.

A whole variety of scope probes is bought and used. Fortunately, it is essential to know which types to use and what scope probe specifications may be chosen.

Types:

- Passive oscilloscope probes.

- Active oscilloscope probes.

Types of Oscilloscope

The following are the different types of oscilloscopes:

- Analogue oscilloscope.

- Analogue storage oscilloscope.

- Digital oscilloscope.

- Digital storage oscilloscope.

- Digital phosphor oscilloscope.

- Digital sampling oscilloscope.

Cathode Ray Tube (CRT)

The CRT is the heart of the CRO. The CRT generates the electron beam, accelerates the beam, deflects the beam and has a screen where beam is visible as a spot. The main parts of the CRT are as follows:

- Electron gun.

- Deflection system.

- Fluorescent screen.

- Glass tube of envelope.

- Base.

Electron Gun

Electron gun section of the CRT provides a sharply focused electron beam which is directed towards the fluorescent coated screen. This screen starts from the thermally heated cathode and electrons are emitted. A negative potential is given to the control grid with respect to cathode. This grid controls number of electrons in the beam going to the screen.

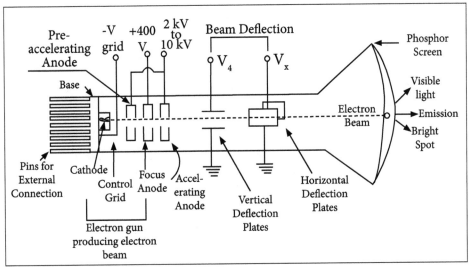

Cathode ray tube.

The momentum of the electrons determines the intensity or brightness of the light emitted from the fluorescent screen due to the electron bombardment. The emitted light is green in colour. Since the electrons are negatively charged, a repulsive force is created by applying negative voltage to the control grid. This negative control voltage is variable.

Since the electron beam consists of many electrons, the beam tends to diverge. This is because, similar charges on the electron repels each other. To compensate for such repulsion forces, an adjustable electrostatic field is created between two cylindrical anodes which result in the acceleration of the electrons.

Both focusing and the accelerating anodes are cylindrical in shape where small openings are located in the center of each electrode, coaxial with the tube axis. The pre-accelerating and accelerating anodes are connected to a common positive high voltage which varies between 2 kV to 10 kV. The focusing anode is connected to a lower positive voltage of about 400 V to 500 V.

Delay Line

It is used to delay the signal for a period of time in the vertical section of CRT. The input signal is not applied directly to the vertical plates because the part of the signal gets lost when the delay time is not used. Therefore, the input signal is delayed by a period of time.

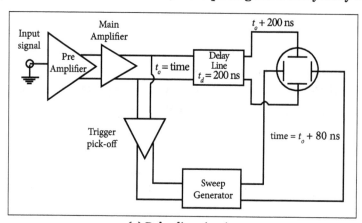

(a) Delay line circuit.

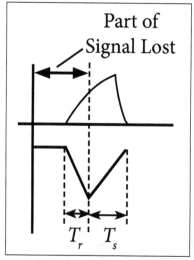

(b) Delay line waveform.

The above diagram shows that, when the delay line is not used, the initial part of the signal is lost and only part of the signal is displayed. To overcome this disadvantage, the signal is not applied directly to the vertical plates but it is passed through a delay line circuit as shown in the above figure (a). This gives time for the sweep to start at the horizontal plates before the signal has reached the vertical plates.

The trigger pulse is picked off at a tune t_o after the signal has passed through the main amplifier. The sweep generator delivers the sweep t_i where the horizontal amplifier and the sweep starts at the HDP at time to + 80 ns. Hence the sweep starts well in time since the signal arrives at the VDP at time t_o + 200 ns.

Deflection System

When the electron beam is accelerated, it passes through the deflection system with which beam can be positioned anywhere on the screen.

The deflection system of the CRT consists of two pairs of parallel plates referred to as the vertical and horizontal deflection plates. One of the plates in each set is connected to ground (0 V). To the other plate of each set, the external deflection voltage is applied through an internal adjustable gain amplifier stage. To apply deflection voltage externally, an external terminal called the Y input or the X-input is available.

As shown in the above figure, the electron beam passes through these plates. A positive voltage applied to the Y input terminal (V_y) causes beam to deflect vertically upward due to the attraction forces while a negative voltage applied to Y input terminal will cause the electron beam to deflect vertically downward due to the repulsion forces.

A positive voltage applied to X-input terminal (V_x) will cause the electron beam to deflect horizontally towards the right while a negative voltage applied to X-input terminal will cause the electron beam to deflect horizontally towards left of the screen. The amount of vertical or horizontal deflection is directly proportional to the corresponding applied voltage.

When the voltage is applied simultaneously to vertical and horizontal deflecting plates, electron beam is deflected due to the resultant of these two voltages. The face of the screen is considered as an x - y plane. The (x, y) position of beam spot is directly influenced by the horizontal and the vertical voltages applied to deflection plates V_x and VY respectively.

The horizontal deflection(X) is proportional to horizontal deflecting voltage (V_x) which is applied to the X-input.

$$K_x = X / V_x$$

Where,

K_x = Constant of proportionality

The deflection produced is measured in cm or number of divisions on the scale in the horizontal direction. Then K_x expressed as cm/volt or division/volt is called as horizontal sensitivity of the oscilloscope.

Vertical deflection(Y) is proportional to the vertical deflecting voltage (V_y) applied to the Y-input.

The values of vertical and horizontal sensitivities are selectable and adjustable through multi positional switches on the front panel that controls the gain of corresponding internal amplifier stage. The bright spot of the electron beam traces x-y relationship between the two voltages Vx and V_y.

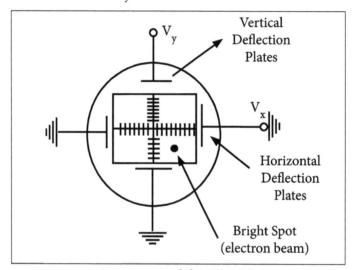

Arrangement of plates in CRT.

Fluorescent Screen

The light produced by screen does not disappear immediately when bombardment by electrons ceases i.e., when the signal becomes zero. The time period for which the trace remains on the screen after the signal becomes zero is known as persistence. The persistence may be as short as a few micro second or as long as tens of seconds or even minutes.

Medium persistence traces are used for general purpose applications. Long persistence traces are used in the study of transients. Long persistence helps in the study of transients since the trace is still seen on the screen after the transient has disappeared. Short persistence is needed for extremely high speed.

The screen is coated with a fluorescent material called phosphor which emits light when bombarded by electrons. There are various phosphors available which differ in colour, persistence and efficiency.

One of the common phosphor is will mite which is combination of zinc, orthosilicate,

$ZnO + SiO_2$ with traces of manganese. This produces the familiar greenish trace. Other useful screen materials include compounds of cadmium, zinc, magnesium and silicon.

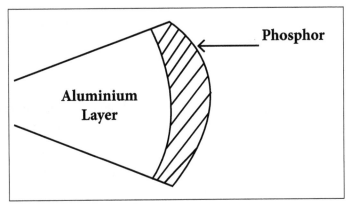

Fluorescent screen.

Kinetic energy of an electron beam is converted into both light and the heat energy when it hits the screen. The heat so produced gives rise to phosphor burn which is destructive. This degrades the light output of phosphor and causes complete destruction of phosphor. Thus phosphor must have high resistance to avoid accidental damage.

The phosphor screen is provided with an aluminium layer called aluminizing. Many phosphor materials have different excitation times and colours as well as different phosphorescence times are available.

Glass Tube: All components of a CRT are enclosed in an evacuated glass tube called as an envelope. This allows the emitted electrons to move freely from one end of the tube to the other end.

Base: The base is provided to the CRT through which the connections are made to various parts.

Deflection Defocussing and its Causes

Whenever an electron beam is deflected from the axial direction, the spot on the fluorescent screen tends to distort and enlarge. This phenomenon due to which spot does not remain in focus and get distorted is called as deflection defocusing.

The various reasons of defocusing are as follows:

- The distance of various points on the screen from the electron gun is not same. The distance from electron gun to screen is greater at the edges due to which defocusing results.

- The non-uniformity in the electric and magnetic deflection fields is used for the deflection. Due to this, part of beam passing through stronger fields are more deflected and part passing through weaker field is less deflected which results in defocussing.

- All the electrons in a beam cannot have exactly same velocity. So due to unequal velocities of the electrons in the beam, defocusing occurs.

Digital CRO

The digital storage oscilloscope replaces the unreliable storage methods used in analog storage scopes with the digital storage with the help of memory. The memory can store data as long as required without degradation.

In this, the waveform to be stored is digitized and then stored in a digital memory. The power to the memory is small and hence stored image can be displayed indefinitely. Once the waveform is digitized, then it can be further loaded into the computer and analysed.

The disadvantages of analog storage cathode ray tube are as follows,

- The waveform preserved for finite amount of time is lost.

- As long as image is required to be stored, the power must be supplied to the tube.

- The trace obtained from the storage tube is not fine when compared to the conventional oscilloscope tube.

- The writing rate of storage tube is less than that of conventional cathode ray tube. This limits the speed of the storage tube.

- The storage cathode ray tube is very expensive than conventional cathode ray tube.

- The storage cathode ray tube requires additional power supplies.

- Only one waveform is stored in storage tube.

- The stored waveform cannot be reproduced on the external device like computer.

Block Diagram

The block diagram of digital storage oscilloscope is shown in the below figure. In all the oscilloscopes, the input signal is applied to the amplifier and attenuators section. The oscilloscope uses same type of amplifier and attenuators circuitry as used in the conventional oscilloscopes. The attenuated signal is then applied to the vertical amplifier.

The vertical input after passing through vertical amplifier is digitized by an analog to digital converter to create a data set which is stored in the memory. The data set is processed by the microprocessor and then sent to the display.

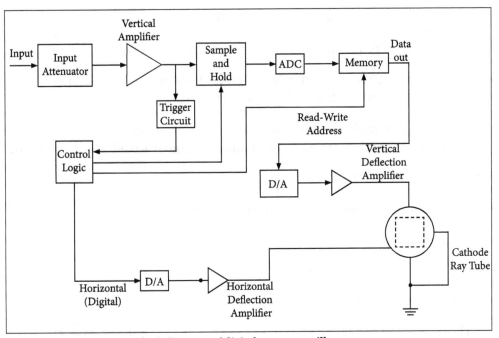

Block diagram of digital storage oscilloscope.

To digitize the analog signal, A/D converter is used. The output of vertical amplifier is applied to the A/D converter section. The main requirement of A/D converter in the digital storage oscilloscope is its speed, while in digital voltmeters, accuracy and resolution are the main requirements. The digitized output needs only binary form and not BCD form. The successive approximation type of A/D converter is used in the digital storage oscilloscope.

The digitizing of the analog input signal takes samples at periodic intervals of the input signal. The rate of sampling should be at least twice as fast as the highest frequency according to sampling theorem. This ensures no loss of information. The sampling rates as high as 100,000 samples per second is used. This requires very fast conversion rate of A/D converter.

If a 12-bit converter is used, 0.025% resolution is obtained while if 10-bit A/D converter is used, then resolution of 0.1 % is obtained. Similarly, with 10-bit A/D converter, the frequency response of 25 kHz is obtained. The total digital memory storage capacity is 4096 for a single channel, 2048 for two channels each and 1024 for four channels each.

The sampling rate and memory size are selected depending on the rate and memory sizes are selected depending on the duration and the waveform to be recorded.

Once the input signal is sampled, the A/D converter digitizes it. The signal is then captured in the memory. Once it is stored in memory, many manipulations are possible as memory can be read out without being erased.

The digital storage oscilloscope has three modes as follows:

- Roll mode - Very fast varying signals are clearly displayed in this mode.

- Store mode - This is called refresh mode. Here, the input initiates the trigger circuits. Memory is refreshed for each triggers.

- Hold or same modes - This is automatic refresh mode. When a save button is pressed, overwriting can be stopped and previously saved signal is locked.

Advantages of digital storage oscilloscope:

- It is easier to operate and has more capability.

- The storage time is infinite.

- The display flexibility is available.

- The cursor measurement is possible.

- The characters can be displayed on screen along with the waveform.

- The X-Y plots, B-H curve, PV diagrams can be displayed.

- The pre-trigger viewing feature allows to display the waveform before trigger pulse.

- Records are kept by transmitting the data to computer system where further processing is possible.

- Signal processing is possible which includes translating the raw data into finished information.

Analogue Storage Scope

It is sometimes necessary to display a signal for a period of time. For signals that have a very long period and the normal persistence of a display, the trace will decay before the whole waveform was complete. A storage facility is required for single shot applications where the single trace is displayed over a period of time to examine the trace.

Analogue storage scopes uses a special cathode ray tube with long persistence facility. A special tube is used with an arrangement to store charge in the area of display where the electron beam had struck which enables the fluorescence to remain for much longer than attainable on normal displays.

Persistence varies in cathode ray tubes, although very bright traces were held over the long periods of time where they permanently burn trace on the screen and these storage displays should be used with care.

5.4 Three Phase Power Measurements and Instrument Transformers

Three Phase Power Measurements

The connection and phasor diagram are shown in the below figure for an assumed abc phase sequence and lagging power factor.

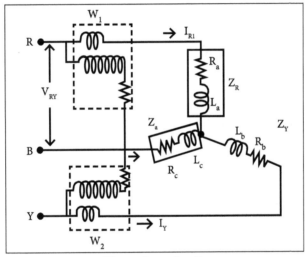

Three phase circuit.

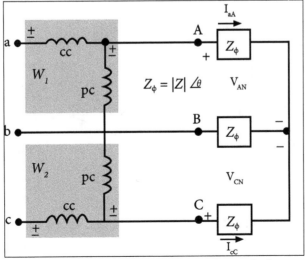

Connection diagram of two-wattmeter method.

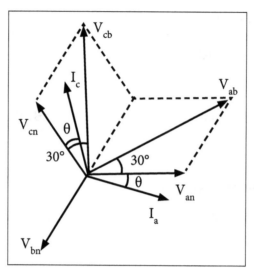

Phasor diagram.

The wattmeter readings are given as,

$$W_1 = V_{ab}I_a\cos\angle(V_{ab}, I_a)$$

$$= V_L I_L \cos(30° + \theta) \quad ...(1)$$

$$W_2 = V_{cb}I_c\cos\angle(V_{cb}, I_c)$$

$$= V_L I_L \cos(30° - \theta) \quad ...(2)$$

The sum of the two wattmeter readings gives the total three phase power,

$$P_T = W_1 + W_2$$

$$= V_L I_L \left[\cos(30° + \theta) + \cos(30° - \theta)\right]$$

$$= \sqrt{3}\, V_L I_L \quad ...(3)$$

The difference of the two wattmeter readings is given as,

$$W_2 - W_1 = V_L I_L \left[\cos(30° - \theta) + \cos(30° + \theta)\right]$$

$$= V_L I_L \sin\theta \quad ...(4)$$

The total reactive power is given as,

$$Q_T = \sqrt{3}\,(W_2 - W_1) \quad ...(5)$$

$$\theta = \tan^{-1}\left(\frac{}{P_T}\right) = \tan^{-1}\left[\frac{\sqrt{3}\,(W_2 \quad W_1)}{W_2 \quad W_1}\right] \qquad ...(6)$$

If $\theta = 0°$, power factor $= 1$ and $W_1 = W_2$

If $\theta = 60°$, power factor $= 0.5$, $W_1 = 0$, $W_2 > 0$

If $\theta = 90°$, power factor $= 0$, $W_1 = -W_2$

One of the watt meters will give negative readings. In the laboratory, when we made the proper wattmeter connections, we will observe that one of the watt meters is trying to read backwards.

After switching the power supply off, the connection of the voltage coil or the current coil (not both) is reversed. The meter will now read upscale. Let us assign a negative sign to this reading.

Advantages of two wattmeter method:

- It is applicable for balanced as well as unbalanced loads.

- Only two watt meters are sufficient to measure total 3 phase power.

- If the load is balanced not only the power but power factor is also determined.

Disadvantages of two wattmeter method:

- It is not applicable for 3 phase, 4 wire system.

- The sign of w_1 and w_2 is identified and noted down correctly. Otherwise, it will lead to the wrong result.

5.4.1 Instrument Transformers (CT and PT)

Instrument transformers are used in AC system for measurement of electrical quantities i.e., voltage, current, power, energy, power factor, frequency. Instrument transformers are also used with protective relays for protection of power system.

Basic function of instrument transformers is to step down the AC system voltage and current. The voltage and current level of power system is very high. It is very difficult and costly to design the measuring instruments for measurement of high level voltage and current. Measuring instruments are designed for 5 A and 110 V.

The measurement of very large electrical quantities is possible by using an instrument transformers with these small rating measuring instruments. Therefore, these instrument transformers are very popular in modern power system.

Instrument transformers.

The requirements of a good PT are:

- Accurate turns ratio: The difficulty in maintaining accurate turns ratio is due to the resistance and reactance of the windings and the value of the exciting current of the transformer.

- Small leakage reactance: The leakage reactance is due to the leakage of magnetic fluxes of the primary and secondary windings. They can be minimized by keeping the primary and secondary windings as close as possible subject to the insulation problem as the primary is at high voltage.

- Small magnetic current: This can be achieved by making the reluctance of the core as small as possible and flux density in the core should also be low, less than 1 Wb/m².

- Minimum voltage drop: The resistance of the windings is made as small as possible. Since primary carries high voltage, it should be heavily insulated. Hence, it is immersed in oil and the terminals are brought out to porcelain bushing. Nowadays, synthetic rubber insulation like styrene is used to avoid oil and porcelain. The theory of PT is the same as power transformer vectorially.

On no load, the secondary voltage is the same as induced voltage. When the load or burden on the secondary is increased (burden is rated as VA), the secondary current increases with corresponding increase in primary current so that the transformation ratio V_p/V_s, remains the same. There is a ratio error due to increased loading but it is of negligible magnitude.

- Phase angle error: As the power factor of the load is reduced from unity, the phase angle error is reduced (lagging). As the power factor of the burden increases, the ratio error increases.

- Effect of frequency: If the frequency is reduced, the flux in the core increases thereby increasing the exciting current and reducing the reactance.

Advantages:

- The normal range voltmeter and ammeter is used along with these transformers to measure high voltage and currents.

- The rating of low range meter can be fixed irrespective of the value of high voltage or current to be measured.

- These transformers isolate the measurement from high voltage and current circuits. This ensures safety of the operator and makes the handling of the equipment very easy and safe.

- It is used for operating many types of protecting devices such as relays or pilot lights.

- Several instruments can be fed economically by single transformer.

Disadvantage:

- The only disadvantage of these instrument transformers is that they can be used only for a.c. circuits and not for d.c. circuits.

Instrument transformers are of two types as follows:

- Current transformer (C.T.).

- Potential transformer (P.T.).

Current Transformer (C.T.)

It is used to step down the current of power system to a lower level to make it feasible to be measured by small rating ammeter (i.e., 5A ammeter). A typical connection diagram of a current transformer is shown in the figure.

Primary of current transformer have a very few turns. Sometimes, bar primary is also used. Primary is connected in series with the power circuit. Therefore, it is also called as series transformer. The secondary have large number of turns. Secondary is connected directly to an ammeter since the ammeter has a very small resistance.

Hence, the secondary of current transformer operates almost in short circuited condition. One terminal of secondary is earthed to avoid the large voltage on secondary with respect to earth which in turn reduces the chances of insulation breakdown and also protects the operator against high voltage. Before disconnecting the ammeter, secondary is short circuited through a switch 'S' as shown in the above figure to avoid the high voltage build up across the secondary.

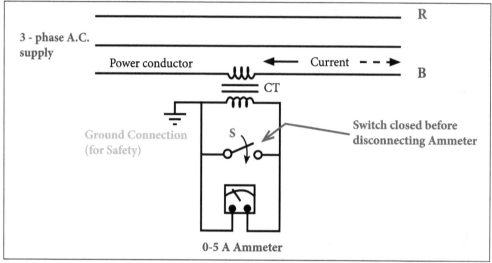

Connection of current transformer.

Characteristics of Current Transformers

The effect of various parameters on the characteristics of current transformers are as follows:

Effect of Power Factor of Secondary Circuit

The power factor of the secondary circuit depends on the power factor of the burden on secondary. This directly affects the two errors of the transformer.

i. Ratio error: For all inductive loads, δ is positive and $\sin(\delta + \alpha)$ is positive. Hence, actual ratio R is always greater than the turns ratio n. For capacitive burdens, δ is negative and R is less than the turn's ratio.

ii. Phase angle error: When the load is inductive and δ is small positive, then θ is positive. When δ approaches 90°, load is highly inductive and θ becomes negative. For capacitive load, δ is negative and θ is always positive.

Effect of Change in I_p

I_p and I_s are directly related where I_p and I_s changes. For low values of I_p, Io is dominating where I_m and I_c are also dominating parts of I_p. Thus errors are higher. When I_p increases, part of Io becomes insignificant from I_p where l_c and I_m are less significant when compared to I_p i.e., Is. Thus, errors are less.

Effect of Change in Burden on Secondary

When the secondary winding circuit burden increases, volt-ampere rating also increases. Due to increased secondary current, secondary flux increases which induces more voltage on secondary. Thus both Im and Ic increases to keep flux constant. Due to this, errors also increase. Thus more secondary burden results in more errors.

Effect of Change in Frequency

If frequency is increased at constant voltage, then the flux and flux density decreases. Thus, there is reduction in I_m and I_c and hence errors are reduced.

Potential Transformer

It is used to step down the voltage of power system to a lower level to make it feasible. A typical connection diagram of a potential transformer is shown in the below figure:

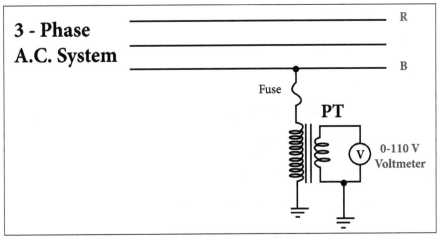

Connection of potential transformer.

Primary of potential transformer have large number of turns. Primary is connected across the line. Hence, it is also called as the parallel transformer. Secondary of potential transformer have few turns and it is connected directly to a voltmeter which have large resistance.

Hence the secondary of a potential transformer operates in an open circuited condition. One terminal of secondary of potential transformer is earthed to maintain the secondary voltage with respect to earth which assures the safety of operators.

Theory of Potential Transformers

The loading of potential transformer is very small and exciting current I_o is in the order of I_s i.e., secondary winding current. In a normal power transformer, I_o is very small compared to I_s.

The equivalent circuit of potential transformer is shown in the below figure:

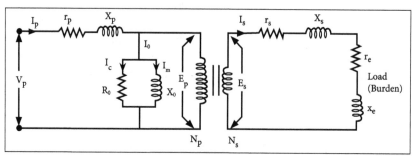

Equivalent circuit of potential transformer.

Where,

ϕ = Working flux

N_p = Primary turns

N_s = Secondary turns

I_p = Primary current

I_s = Secondary current

I_m = Magnetizing component of I_o

l_c = Core less component of l_o

l_o = No load current i.e., exciting current

r_s, x_s = Resistance and reactance of secondary winding

r_p, x_p = Resistance and reactance of primary winding

r_e, x_e = Resistance and reactance of burden

E_p = Primary induced voltage

E_s = Secondary induced voltage

Δ = Phase angle of secondary load current = $\tan^{-1} \dfrac{x_e}{r_e}$

V_p = Primary applied voltage

V_s = Secondary terminal voltage

For potential transformer,

$$n = \frac{N_P}{N_S} = \frac{E_P}{E_S}$$

Phasor diagram is shown in the below figure:

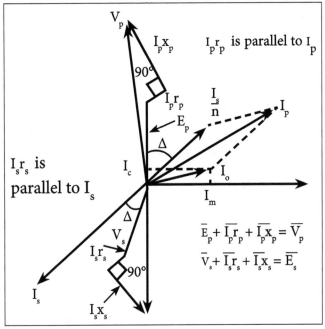

Phasor diagram of a potential transformer.

Derivation of Phase Angle (θ)

$$\tan\theta = \frac{ab}{Oa} = \frac{I_p \times x_p \cos\beta - I_p r_p \sin\beta + n I_s x_s \cos\Delta - n I_s r_s \sin\Delta}{n V_s + n I_s r_s \cos\Delta + n I_s x_p \sin\Delta + I_p r_p \cos\beta + I_p x_p \sin\beta}$$

Since θ is very small, $\tan\theta \approx \theta$

$$\therefore \quad \theta = \frac{I_p \times x_p \cos\beta - I_p r_p \sin\beta + n I_s x_s \cos\Delta - n I_s r_s \sin\Delta}{n V_s}$$

$$= \frac{x_p \left[\dfrac{I_s}{n}\cos\Delta + I_c\right] - r_p \left[I_m + \dfrac{I_s}{n}\sin\Delta\right] + n I_s x_s \cos\Delta - n I_s r_s \sin\Delta}{n V_s}$$

$$= \frac{I_s \cos\Delta\left(\dfrac{x_p}{n} + n x_s\right) - I_s \sin\Delta\left(\dfrac{r_p}{n} + n r_s\right) + I_c x_p - I_m r_p}{n V_s}$$

$$= \frac{\dfrac{I_s \cos\Delta}{n}\left(x_p + n^2 x_s\right) - \dfrac{I_s \sin\Delta}{n}\left(r_p + n^2 r_s\right) + I_c x_p - I_m r_p}{n V_s}$$

$$r_p + n^2 r_s = R_{1e} \quad \text{and} \quad x_p + n^2 x_s = X_{1e}$$

$$\therefore \quad \theta = \frac{\dfrac{I_s}{n}\cos\Delta\, X_{1e} - \dfrac{I_s}{n}\sin\Delta\, R_{1e} + I_c x_p - I_m r_p}{n V_s}$$

$$\therefore \quad \theta = \frac{\dfrac{I_s}{n}\left(X_{1e}\cos\Delta - R_{1e}\sin\Delta\right) + I_c x_p - I_m r_p}{n V_s} \text{radians} \quad ...(1)$$

$$R_{2e} = \frac{R_{1e}}{n^2} \quad \text{and} \quad X_{2e} = \frac{X_{2e}}{n^2}$$

$$\therefore \quad \theta = \frac{n I_s\left(X_{2e}\cos\Delta - R_{2e}\sin\Delta\right) + I_c x_p - I_m r_p}{nV}$$

$$\therefore \quad \theta = \frac{I_s}{V_s}\left(X_{2e}\cos\Delta - R_{2e}\sin\Delta\right) + \frac{I_c x_p - I_m r_p}{n V_s}\text{radians}$$

It is noted that the phase angle θ is treated positive when V_s is reversed i.e., n_{Vs} leads the primary winding voltage V_p. θ is treated negative when n_{Vs} lags the primary winding voltage V_p.

Once R and θ are obtained, then the errors in potential transformers are given as,

$$\% \text{ ratio error } = \frac{K_n - R}{R}\times 100$$

Phase angel error $= \theta$ radians

Ratio Error

Current transformation ratio I_2/I_1 is equal to the turns ratio N_1/N_2. The current ratio is not equal to turns ratio due to the magnetizing and core loss components of the exciting current. It is affected due to the secondary current and its power factor.

The load current is not a constant fraction of the primary current. In case of potential transformers, the voltage ratio V_2/V_1 is not exactly equal to N_2/N_1.

Transformation ratio is not constant but depends on the load current, power factor of the load and exciting current of the transformer. Due to this fact, a large error is introduced in the measurements done by the instrument transformers. Such an error is referred to as ratio error.

The ratio error is defined as,

$$\% \text{ Ratio error} = \frac{\text{Nominal ratio} - \text{Actual ratio}}{\text{Actual ratio}} \times 100$$

$$\% \text{ Ratio error} = \frac{K_n - R}{R} \times 100$$

Phase Angle Error

In the power measurements, phase of a secondary current is displaced by exactly $180°$ from that of primary current for current transformer. The phase of the secondary voltage is displaced exactly $180°$ from that of primary voltage for primary transformer. It denoted by an angle θ where the phase difference between the primary and secondary is different from $180°$.

Phase angle error is given as,

$$\theta = \frac{180}{\pi} \left[\frac{I_m \cos \delta - I_c \sin \delta}{n I_s} \right] \text{ degrees}$$

Loads are inductive and δ is positive and very small.

$$\sin \delta = 0$$

$$\cos \delta = 1$$

$$R = n + \frac{I_c}{I_s}$$

$$\theta = \left(\frac{180}{\pi} \right) \left(\frac{I_m}{n I_s} \right) \text{ degrees}$$

$$n = \frac{I_p}{I_s}$$

$$R = n + \frac{n I_c}{I_p}$$

$$\theta = \left(\frac{180}{\pi} \right) \left(\frac{I_m}{I_p} \right) \text{ degrees}.$$

Differences between C.T. and P.T

SI. No	Current transformer (C.T.)	Potential transformer (P.T.)
1.	It is connected in series with power circuit.	It is connected in parallel with power circuit.
2.	Secondary is connected to ammeter.	Secondary is connected to voltmeter.
3.	Secondary works almost in short circuited condition.	Secondary works almost in open circuited condition.
4.	Primary current depends on power circuit current.	Primary current depends on secondary burden.
5.	Primary current and excitation vary over a wide range with change of power circuit current.	Primary current and excitation variation are restricted to a small range.
6.	One terminal of secondary is earthed to avoid the insulation break down.	One terminal of secondary is earthed for safety.
7.	Secondary is never open circuited.	Secondary can be used in open circuit condition.

Permissions

We would like to thank the editorial team for lending their expertise to make the book truly unique. They have played a crucial role in the development of this book. Without their invaluable contributions this book wouldn't have been possible. They have made vital efforts to compile up to date information on the varied aspects of this subject to make this book a valuable addition to the collection of many professionals and students.

This book was conceptualized with the vision of imparting up-to-date and integrated information in this field. To ensure the same, a matchless editorial board was set up. Every individual on the board went through rigorous rounds of assessment to prove their worth. After which they invested a large part of their time researching and compiling the most relevant data for our readers.

The editorial board has been involved in producing this book since its inception. They have spent rigorous hours researching and exploring the diverse topics which have resulted in the successful publishing of this book. They have passed on their knowledge of decades through this book. To expedite this challenging task, the publisher supported the team at every step. A small team of assistant editors was also appointed to further simplify the editing procedure and attain best results for the readers.

Apart from the editorial board, the designing team has also invested a significant amount of their time in understanding the subject and creating the most relevant covers. They scrutinized every image to scout for the most suitable representation of the subject and create an appropriate cover for the book.

The publishing team has been an ardent support to the editorial, designing and production team. Their endless efforts to recruit the best for this project, has resulted in the accomplishment of this book. They are a veteran in the field of academics and their pool of knowledge is as vast as their experience in printing. Their expertise and guidance has proved useful at every step. Their uncompromising quality standards have made this book an exceptional effort. Their encouragement from time to time has been an inspiration for everyone.

The publisher and the editorial board hope that this book will prove to be a valuable piece of knowledge for students, practitioners and scholars across the globe.

Index